T0224505

SpringerBriefs in Computer Science

SpringerBriefs present concise summaries of cutting-edge research and practical applications across a wide spectrum of fields. Featuring compact volumes of 50 to 125 pages, the series covers a range of content from professional to academic.

Typical topics might include:

- A timely report of state-of-the art analytical techniques
- A bridge between new research results, as published in journal articles, and a contextual literature review
- A snapshot of a hot or emerging topic
- An in-depth case study or clinical example
- A presentation of core concepts that students must understand in order to make independent contributions

Briefs allow authors to present their ideas and readers to absorb them with minimal time investment. Briefs will be published as part of Springer's eBook collection, with millions of users worldwide. In addition, Briefs will be available for individual print and electronic purchase. Briefs are characterized by fast, global electronic dissemination, standard publishing contracts, easy-to-use manuscript preparation and formatting guidelines, and expedited production schedules. We aim for publication 8–12 weeks after acceptance. Both solicited and unsolicited manuscripts are considered for publication in this series.

More information about this series at http://www.springer.com/series/10028

Jie Hu · Kun Yang

Data and Energy Integrated Communication Networks

A Brief Introduction

Jie Hu
School of Information and Communication Engineering
University of Electronic Science and Technology of China
Chengdu, Sichuan
China

Kun Yang
School of Computer Science and Electronic Engineering
University of Essex
Colchester
UK

ISSN 2191-5768 ISSN 2191-5776 (electronic)
SpringerBriefs in Computer Science
ISBN 978-981-13-0115-5 ISBN 978-981-13-0116-2 (eBook)
https://doi.org/10.1007/978-981-13-0116-2

Library of Congress Control Number: 2018943388

Printed on acid-free paper

This Springer imprint is published by the registered company Springer Nature Singapore Pte Ltd.
The registered company address is: 152 Beach Road, #21-01/04 Gateway East, Singapore 189721, Singapore

Preface

In order to satisfy the power thirsty of communication devices in the imminent fifth-generation (5G) and Internet of Things (IoT) era, wireless charging techniques have attracted much attention both from the academic and industrial communities. Although the inductive coupling and magnetic resonance based charging techniques are indeed capable of supplying energy in a wireless manner, they tend to restrict the freedom of movement. By contrast, RF signals are capable of supplying energy over distances, which are gradually inclining closer to our ultimate goal—charging anytime and anywhere. Furthermore, transmitters capable of emitting RF signals have been widely deployed, in TV towers, cellular base stations and WiFi access points. This communication infrastructure may indeed be employed also for wireless energy transfer (WET). Therefore, no extra investment in a dedicated WET infrastructure is required. However, allowing radio frequency (RF) signal based wireless energy transfer (WET) may impair the wireless information transfer (WIT) operating in the same spectrum. Hence, it is crucial to coordinate and balance WET and WIT for simultaneous wireless information and power transfer (SWIPT), which evolves to data and energy integrated communication networks (DEINs). This brief aims for providing a landscape picture of DEINs, while including latest research contributions in this promising topic.

To this end, we first provide an overview of DEIN in Chap. 1. We will look into the energy shortage of the electronic devices, compare the popular wireless charging techniques with one another and highlight the RF signal based WET and its distinctive features against the conventional wireless communication in the same spectral bands. Then, we will describe the ubiquitous architecture of DEINs.

In Chap. 2, we will focus on the fundamental of the physical layer for implementing the integrated WET and WIT of the point-to-point link. Key enabling modules of the generic transceiver architecture for the integrated WET and WIT will be introduced. Then, we will introduce several popular receivers equipped with multiple antennas for simultaneously information and energy reception, namely the ideal receiver, the spatial splitting based receiver, the power splitting based receiver and the time switching based receiver.

In Chap. 3, we consider a typical DEIN system consisting of a single H-BS and multiple DEIN users, who are eager to receive both information and energy simultaneously during the downlink transmission of the H-BS. The DEIN users then exploit the energy harvested from the downlink for powering their own uplink transmission. Both the downlink and uplink transmissions are time slotted in order to reduce the potential interference and transmission collision among the multiple users. Optimal time slot allocation schemes in the MAC layer are proposed for maximising the sum-throughput and the fair-throughput of the DEIN users' uplink transmissions, respectively.

In Chap. 4, a full-duplex aided H-BS is conceived in a multi-user DEIN for the sake of simultaneously transferring energy during its downlink transmission, while receiving the information uploaded by the multiple users. The uplink transmissions of the multiple users are powered by the energy harvested from the H-BS's downlink energy broadcast. In this full-duplex aided DEIN, a joint time allocation and user scheduling algorithm is proposed for the sake of maximising the sum-throughput of the users' uplink transmissions by further considering the users' actual information uploading requirements.

Finally, we conclude this brief in Chap. 5 by providing some emerging research topics in the DEIN.

This brief aims for boosting the joint effort from both the academia and industry so as to push the DEIN a step closer to the practical implementation. It is also suitable for the undergraduate/postgraduate students to be familiar with this cutting-edge technique.

We would like to thank Mr. Yizhe Zhao and Mr. Kesi Lv for their tremendous contribution to Chaps. 2–5. We would also like to thank Prof. Xuemin (Sherman) Shen, University of Waterloo, for his outstanding editorial organisation of this influential series in the computer science. The financial support of the National Natural Science Foundation of China (NSFC), Grant No. 61601097, U1705263, and 61620106011, as well as that of Fundamental Research Funds for the Central Universities, Grant No. ZYGX2016Z011 are gratefully acknowledged. This work is also sponsored by Huawei Innovative Research Program (HIRP).

Chengdu, China Jie Hu
Colchester, UK Kun Yang

Contents

Acronyms

AC	Alternative Current
AWGN	Additive White Gaussian Noise
CDF	Cumulative Distribution Function
CDMA	Code Division Multiple Access
CICO-MC	Continuous Input Continuous Output Memoryless Channel
DC	Direct Current
DEIN	Data and Energy Integrated communication Network
DIDO-MC	Discrete Input Discrete Output Memoryless Channel
FDD	Frequency Division Duplex
H-BS	Hybrid Base Station
IoT	Internet of Things
KKT	Karush–Kuhn–Tucker
LPF	Low-Pass Filter
MAC	Medium Access Control
MIMO	Multiple Input Multiple Output
MISO	Multiple Input Single Output
mmW	millimetre Wave
NASA	National Aeronautics and Space Administration
NOMA	Non-Orthogonal Multiple Access
NSFC	National Natural Science Foundation of China
OFDMA	Orthogonal Frequency Division Multiple Access
PAPR	Peak to Average Power Ratio
PS	Power Splitting
PSK	Phase Shift Keying
QAM	Quadrature Amplitude Modulation
QoS	Quality of Service
RF	Radio Frequency
SCMA	Sparse Code Multiple Access
SER	Symbol Error Ratio
SIMO	Single-Input-Multiple-Output

SISO	Signle-Input-Single-Output
SS	Spatial Splitting
SVD	Singular Value Decomposition
SWIPT	Simultaneous Wireless Information and Power Transfer
TDD	Time Division Duplex
TDMA	Time Division Multiple Access
TS	Time Switching
UE	User Equipment
UESTC	University of Electronic Science and Technology of China
WET	Wireless Energy Transfer
WIT	Wireless Information Transfer
WPCN	Wireless Powered Communication Network
5G	Fifth Generation

Chapter 1
Data and Energy Integrated Communication Networks: An Overview

Abstract In order to address the energy supply issue of communication devices in the imminent 5G and IoT era, wireless charging techniques have attracted much attention both from the academic and industrial communities. Thankfully, RF signals are capable of delivering energy over distances. However, allowing RF signal based wireless energy transfer (WET) may impair the wireless information transfer (WIT) operating in the same spectral band. Hence, it is crucial to coordinate and balance WET and WIT for simultaneous wireless information and power transfer (SWIPT), which evolves to Data and Energy Integrated communication Networks (DEINs). To this end, a ubiquitous IDEN architecture is characterised by summarising its natural heterogeneity and by synthesising a diverse range of integrated WET and WIT scenarios.

Keywords Data and Energy Integrated Communication Network (DEIN)
Energy Efficiency · RF Signal based Wireless Charging · Simultaneous Wireless Information and Energy Transfer (SWIPT) · Ubiquitous Architecture of DEIN
Wireless Energy Transfer (WET) · Wireless Information Transfer (WIT)
Wireless Powered Communication Network (WPCN)

We provide an overview of Data and Energy Integrated communication Network (DEIN) in this chapter. We will look into the energy shortage of the electronic devices, compare the popular wireless charging techniques with one another and highlight the RF signal based Wireless Energy Transfer (WET) and its distinctive features against the conventional wireless Information Transfer (WIT) in the same spectral bands. Then we will describe the ubiquitous architecture of DEINs by introducing its natural heterogeneity and by synthesising a diverse range of WET and WIT scenarios.

1.1 Energy Dilemma for Electronic Devices

According to the prediction of the classic Moore's Law, the density of transistors in an integrated circuit doubles approximately every two years, which have been fuelling the spectacular proliferation of electronic devices since the 1960s. Furthermore, con-

J. Hu and K. Yang, *Data and Energy Integrated Communication Networks*, SpringerBriefs in Computer Science, https://doi.org/10.1007/978-981-13-0116-2_1

sumer electronic devices are becoming shirt-pocket-sized and mobile. These devices are normally powered by embedded batteries. However, as their functions become ever more sophisticated, their thirst for abundant energy is not matched by the slow progress of the batteries' capacity. The situation in the communication industry is even more daunting. Since the roll-out of the fifth-generation (5G) cellular system and of the Internet of Things (IoT) is just around the corner, people's appetite for super-high data transmission rates, for high density of connectivity and for high mobilities will indeed be satisfied to a large extent. A major portion of the future mobile data traffic will be constituted by novel types of services, including high-definition stero-scopic video streams, augmented/virtual reality, holographic tele-presence, cloud desktops, as well as online games, etc. All these services require the user terminals to be implemented with high computing capabilities for real-time signal processing, which may quickly drain the embedded batteries. Furthermore, sensors will be deployed in every corner of the future smart cities [1]. These sensors monitor the environment and upload sensing results to central servers [2]. The life-span of sensors and of sensing networks largely depends on the sensors' battery capacity. Regularly replacing the batteries may be an unrealistic or tedious task. Accordingly, new sources of energy have to be explored to prolong the depletion period of conventional batteries in order to relieve the energy concerns of various communication devices.

1.2 Near-Field Wireless Energy Transfer

Nowadays, resonant inductive coupling [3] and magnetic resonance coupling [4] have emerged for remotely charging electronic devices in the near-field. Resonant inductive coupling based wireless charging relies on the magnetic coupling that delivers electrical energy between two coils tuned to resonate at the same frequency. This technique has already been commercialised for some home electronic appliances [5], such as mobile phones, electric toothbrushes and smart watches etc. However, the coupling coils only support near-field wireless energy transfer (WET) over a distance spanning from a few millimetres to a few centimetres [6], while achieving a WET efficiency as high as 56.7%, when operating at a frequency of 508 kHz [7]. Furthermore, resonant inductive coupling requires strict alignment of the coupling coils. Even a small misalignment may result in dramatic reduction of the WET efficiency [8]. As a result, during the charging process, the electronic appliances cannot be freely moved.

By contrast, magnetic resonance coupling [9] delivers electrical energy between two resonators by exploiting evanescent-wave coupling. This technique has already been adopted for charging the electric vehicles due to its high WET efficiency [10]. For example, magnetic resonance coupling is capable of achieving a WET efficiency of 90% over a distance of 0.75 m [11]. Both its WET efficiency and its charging distance are much higher than that of the resonant inductive coupling. However, magnetic resonance coupling still belongs to the category of near-field wireless charging, since its power transfer efficiency dramatically reduces to 30%, when the distance is increased to 2.25 m [11]. Nonetheless, magnetic resonance coupling does not require strict alignment between the rechargeable device and the energy source. Hence, during the charging process, the electronic appliances may be moved within the charging area [12]. Furthermore, a multiple-input-multiple-output (MIMO) system, which has already been widely adopted for improving the performance of the wireless communication, can also be introduced into the magnetic resonance coupling based WET system for the sake of further enhancing the WET efficiency [13, 14].

1.3 RF Signal Based Wireless Energy Transfer

In contrast to the above-mentioned near-field WET techniques, the propagation of the RF signals is capable of supporting far-field WET [15]. The history of the RF signal based WET dates back to 1960, when the first long-distance WET system was established by Brown [16, 17]. Brown jointly designed rectifiers and antennas for energy receivers, which is now widely known as rectennas. They are capable of efficiently converting the Alternating-Current (AC) energy carried by the RF signals to Direct Current (DC) energy. This RF signal based WET system was validated by remotely powering a model helicopter from the ground in 1964 [16, 17]. In the 1970s and 1980s, intense efforts were invested into the research of RF signal based WET, which was largely motivated by the intention of developing a solar-powered satellite [18, 19]. In this system, a satellite may harvest energy from sunlight in the outer space and beam the energy back to ground stations via the propagation of RF signals. Furthermore, the Jet Propulsion Laboratory of the National Aeronautics and Space Administration (NASA) led a project from 1969 to 1975, in which 30 kW of power was beamed over a distance of 1 mile at a 84% RF-DC efficiency [20].

There are three main technical challenges in the RF based WET. Firstly, the long-distance propagation and adverse multipath fading may substantially attenuate the RF signals before they arrive at the receivers, which inevitably results in energy loss. Secondly, the energy carried by RF signals is of AC nature, which cannot be directly invoked for driving an electronic load. As a result, the AC energy carried by RF signals have to be converted to DC energy for any further use. However, some portion of energy is inevitably lost during this conversion process. Last but not the least, the diffraction of the RF signals' waveform may expand the beam size. As a result, the receive antenna having limited size is not capable of capturing all the energy carried by the RF signals. For counteracting the signal attenuation of wireless channels, the transmit beams have to be accurately aimed at the energy receivers [21], which requires the joint design of the transmit and receive antennas. For improving the AC-DC conversion efficiency, the receive antennas have to be designed together with the rectifiers in order to achieve the impedance match for the sake of high-efficiency AC-DC conversion [22]. For alleviating the adverse effect of beam diffraction, the non-diffracted Bessel-Gaussian beam [23] can be invoked, which is capable of efficiently reducing the energy loss during the propagation and hence improve the WET efficiency over wireless channels.

In general, the RF signal based WET has the following advantages over its near-field counterparts:

- Large coverage. Relying on the RF signals, energy can be transferred to receivers miles away.
- High flexibility. The angular selectivity transmit beam can be intelligently adjusted according to various WET requirements. For instance, a narrow beam can be invoked for realising accurate and high-efficiency point-to-point WET, while a wide beam can be used for simultaneously charging multiple devices.
- More applications. RF signals can be leveraged for supplying a large amount of energy to energy-hungry appliances, such as solar-powered satellite system. It can also be exploited for supplying energy to low-power devices, such as sensors and biomedical implants.
- Low investment. The transmitters of the RF signals have been deployed at every corner of the globe, such as radio broadcast stations, TV towers, cellular base stations and WiFi access points, etc. The legacy of the communication infrastructure can all be exploited for radiating energy to electronic devices. Only limited extra investment is required for deploying energy transmitters in order to cover some blind spots.

The main features of different WET techniques are summarised in Table 1.1.

Table 1.1 Main features of different WET techniques

Technique	Range	Direc.	Frequency	Antenna	Application
RF signals	Long	High	MHz-GHz	Parabolic dishes, rectennas, phased arrays	Solar-powered satellite, drone aircraft, IoT devices, portable devices, RFID, smart cards and etc.
Magnetic resonant coupling	Middle	Low	kHz-GHz	Tuned wire coils, lumped element resonators	Portable devices, biomedical implants, electric vehicles, RFID, smartcard and etc.
Inductive coupling	Short	Low	Hz-MHz	Wire coils	Stovetops, industrial heaters and small electric appliances, such as electric toothbrush, razor and etc.

1.4 WET Versus WIT in the RF Spectral Band

Since RF signal based WET techniques require highly fexible beam directivity in order to satisfy diverse charging requests, the best spectral band for steering energy beams is in the range of 10 MHz to 100 GHz, which almost covers all the bands allocated for wireless communication services. For example, TV/Radio broadcasting services operate in the band spanning from 40 MHz to 220 MHz [24], the mobile cellular communication system operates in the spectral band spanning from 800 MHz to 3.7 GHz [25], while the WiFi communication system operates in the spectral band spanning from 2.4 GHz to 6 GHz [26]. Furthermore, as a key technique in the upcoming 5G era, millimetre wave (mmW) [27] may significantly increase the achievable throughput of the air interface, which operates in the spectral band ranging from 10 GHz to 100 GHz.

Although they both operate in the same RF band, WET and WIT still have the following distinctive characteristics:

- They have different functional circuits. RF signals in the pass-band cannot be directly invoked for both the information decoding and the energy harvesting. For the information decoding, the RF signals in the pass-band have to be firstly converted to the base-band, since all the signal processing has to be accomplished in the base-band. By contrast, for the energy harvesting, the AC energy carried by the RF signals has to be converted to the DC energy first, since only DC energy can be stored in batteries or drive electronic loads. Specifically, during the AC-DC conversion, the phase information carried by the RF signals is filtered.
- They require different absolute energy at receivers. The activation of the energy harvesting circuits requires a relatively high energy carried by the received RF signals, which is approximately on the order of −20 dBm. If the energy carried by the received RF signal does not achieve the required activation threshold, none of this energy can be harvested. By contrast, the successful information recovery relies on the energy ratio between the received RF signal and the noise plus interference, not on the absolute energy carried by the received RF signal. As a result, even a small amount of energy is capable of activating the information receiver, which is approximately on the order of −80 dBm.
- They have different coverage. The RF signals are attenuated by hostile wireless channels, such as the path loss, shadowing and multipath fading. Since the energy harvesting requires a much higher absolute energy at the receivers than the information decoding, the range of WET is accordingly much shorter than that of WIT. Therefore, given the same set of transmitters and receivers, the resultant WET network has a different topology with the WIT network.
- They treat noise and interference differently. The interference and noise ubiquitously exist in any WIT system, which seriously impair the WIT performance. Mitigating the performance degradation induced by the interference and noise is a major challenge in the WIT system design. By contrast, WET systems may actually benefit from the interference and noise, since both of them are RF signals and they both carry useful energy. The interference and noise can be jointly harvested

by the energy harvesting circuits, which may provide additional energy harvesting gains for the energy requesters.

- They have different definitions in energy efficiency. The energy efficiency of WET can be defined as the ratio of energy harvested by the receiver to the energy emitted by the transmitter, which can be formulated as

$$\eta_{WET} = \frac{1}{P_t} \cdot \rho\,(P_r + P_I + P_N) \quad \text{(Watt/Watt)}, \tag{1.1}$$

where ρ is the conversion rate from the received RF energy to the DC energy by considering a linear RF-DC converter. By contrast, In the community of green communications, the energy efficiency of WIT is defined as the ratio of spectral efficiency to energy consumption, which is evaluated in the unit of bps/Hz/Watt or bps/Hz/Joule. By exploiting the classic Shannon-Hartley theorem in an Additive-White-Gaussian-Noise (AWGN) channel, the energy efficiency of WIT can be expressed as

$$\eta_{WIT} = \frac{1}{P_t} \cdot \log_2\left(1 + \frac{P_r}{P_I + P_N}\right) \quad \text{(bps/Hz/Watt)}, \tag{1.2}$$

where P_t is the transmit power of the RF signal, P_r is the power received after the signal being attenuated by the hostile wireless channel, P_I is the aggregate interference power and P_N is the noise power at the receiver.

In Fig. 1.1, we exemplify the energy efficiency of WET and that of WIT, which can be calculated by (1.2) and (1.1), respectively. Observe from Fig. 1.1a that in our setting, the energy efficiency of WET reduces from 1.1% but converges to 1%, which is due to the channel attenuation incurred by the path loss between the transmitter and receiver pair. Observe from Fig. 1.1b that the energy efficiency of WIT gradually reduces from 35 bps/Hz/mW to 0 as the transmit power of the RF signal increases. By contrast, WET and WIT operating in the same RF spectral band may compete for the precious resources in the air interface and they may thus impair each other's performance to some extent. For example, WET requires that the RF signals carry a high power to the receivers for the efficient energy harvesting. However, the high-power RF signals of the WET system may impose excessive interference on the WIT receivers, which may thus significantly degrade the WIT performance attained. As a result, coordinating WET and WIT in the same RF band imposes critical challenges on the RF circuit design, on the transceiver design of the physical layer, on the resource scheduling/allocation schemes and on the corresponding protocol design of the medium-access-control (MAC) layer. Furthermore, integrated data and energy transfer in the RF band also requires a joint networking concept for heterogeneous data and energy transceivers. All these challenging issues require novel Data and Energy Integrated Communication Networks (DEINs) [33].

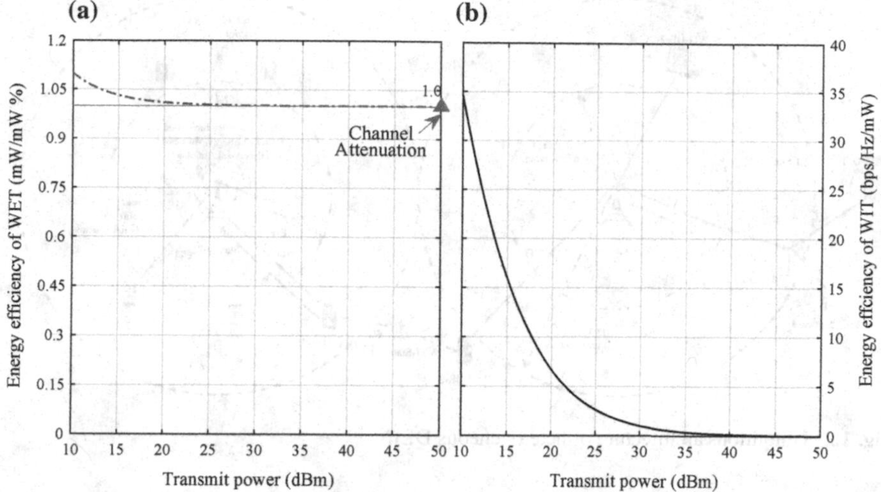

Fig. 1.1 Energy efficiency of WET (**a**) and that of WIT (**b**) against transmit power of RF signals. The noise power is $P_N = -94$ dBm, which is calculated by the power spectrum density of the thermal noise −174 dBm/Hz and 100 MHz of the RF signals' bandwidth. The aggregate interference power at the receiver is set to be $P_I = -20$ dBm, which appears in a heterogeneous cellular network with the highest probability [28]. The distance between a transmitter and receiver pair is 10 m. The path loss is calculated by the model invoked in [29–32], where the path loss exponent is 2. No fading is assumed. The antenna gain in this example is set to be 40 dBi in order to counteract the path loss

1.5 Ubiquitous Architecture of the DEIN

DEINs are naturally heterogeneous in terms of all their technical aspects. We will investigate the heterogeneity of the DEINs and synthesise a diverse range of WET and WIT scenarios into its ubiquitous architecture, which is exemplified in Fig. 1.2.

1.5.1 Heterogeneous Infrastructure

First of all, there are various types of infrastructure elements in heterogeneous DEIN. As portrayed in Fig. 1.2, we have generally three basic type of infrastructure in DEINs, namely DEIN stations, WET stations and WIT stations/relays. DEIN stations [34] are capable of operating both as information transmitter and as energy transmitter for satisfying both of the user equipments' (UEs') data and energy requests. Thanks to their powerful functionalities, DEIN stations are also capable of realising integrated data and energy transfer for the sake of increasing the spectrum efficiency of the congested RF band. Therefore, DEIN stations have to be connected to the core communication network and they also have to be powered by stable energy sources, such as large solar energy harvesters and the power grid. As illustrated in Fig. 1.2,

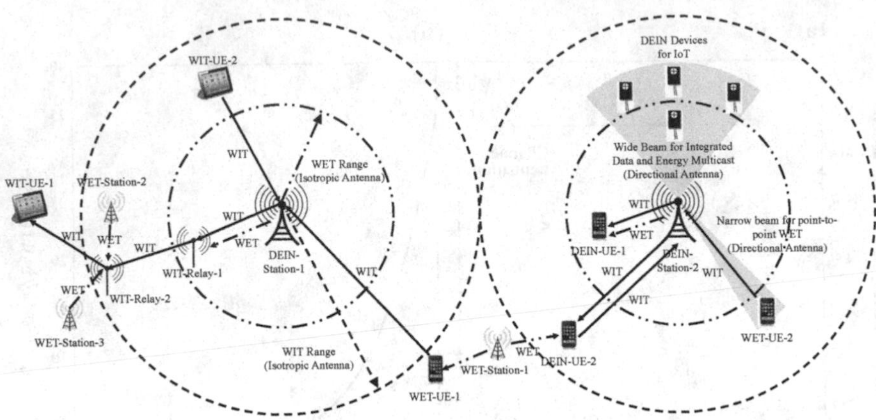

Fig. 1.2 Ubiquitous architecture of heterogeneous DEIN

DEIN-Station-1 may satisfy the integrated data and energy requests from the IoT devices and those from DEIN-UE-1.

However, as we have discussed in Sect. 1.4, the reliable WET range is far shorter than the reliable WIT range, as exemplified in Fig. 1.2. As a result, some blind areas cannot be adequately covered by WET of DEIN stations. Furthermore, some dedicated WET [35] stations are also deployed in order to supply energy to the devices roaming in these blind areas. These WET stations are only connected to energy sources, but they do not have to be connected to the core communication network. As a result, they are dedicated for satisfying the UEs' charging requests. For instance, as shown in Fig. 1.2, three WET stations are deployed in order to supply energy to the UEs beyond the WET range of the DEIN stations.

Apart from DEIN stations and WET stations, there are still many conventional communication stations in heterogeneous DEINs, namely the classic femto-cellular stations, pico-cellular stations and macro-cellular stations [36]. These communication stations have different levels of transmit power and coverage, which results in obvious heterogeneity in DEINs. Sometimes, low-cost relay stations are also deployed for forwarding the data packets to cell-edge UEs, as illustrated in Fig. 1.2. However, small cellular stations and relay stations [37] are only capable of emitting RF signals at a limited power. They are not suitable for carrying out sophisticated WET tasks. Therefore, they are regarded as a dedicated communication infrastructure.

1.5.2 Heterogeneous User Equipment

Apart from the heterogeneous infrastructure, our DEINs have to accommodate both charging and communication requests from diverse types of UEs. We generally have three types of UEs in DEINs, namely the WIT UEs, the WET UEs and the DEIN UEs [38], as exemplified in Fig. 1.2. WIT UEs only require downlink and uplink data transmission in DEINs. Since these UEs are always powered by stable energy sources, they do not request any wireless charging from the DEIN stations. Laptops and tablets are typical WIT UEs, which are either powered by high-capacity batteries or are connected to the power grid. For example, as illustrated in the left part of Fig. 1.2, WIT-UE-1 receives its requested data from DEIN-Station-1 with the aid of two WIT relay stations, while WIT-UE-2 may consume its own energy for powering its uplink information transmission.

By contrast, since WET UEs are not powered by stable energy sources, they have to request additional energy supply either from the DEIN stations or from the WET stations in order to support their basic functionalities, such as uplink information transmissions and energy-consuming computations [39]. For instance, although WET-UE-1 is beyond the WIT range of DEIN-Station-1, it may still establish reliable uplink transmissions with DEIN-Station-1 by exploiting the additional energy received from WET-Station-1, as exemplified in Fig. 1.2. Similarly, the uplink transmission of WET-UE-2 towards DEIN-Station-2 is powered by DEIN-Station-2 itself. Miniature-sized IoT devices are typical WET UEs, since their functionalities are limited by the amount of energy stored in their batteries.

Furthermore, some UEs simultaneously request data and energy transmissions, which are regarded as DEIN UEs [40]. For instance, in the right cell of Fig. 1.2, DEIN-UE-1 simultaneously receives its requested data and energy from DEIN-Station-2, while DEIN-UE-2 also simultaneously requests both downlink data transmission and wireless charging. However, since DEIN-UE-2 is beyond the WET range of DEIN-Station-1, it can only receive the requested data from DEIN-Station-2, but it can receive energy from WET-Station-1. This energy may be exploited for supporting DEIN-UE-2's uplink data transmission to its associated DEIN-Station-2.

Sometimes, the functionalities of WIT relay stations are also limited by their energy supply, especially when the WIT relay stations rely on energy gleaned from batteries or harvested from renewable sources. As a result, they also need wireless charging from DEIN stations or WET stations for powering their data packet forwarding actions [41]. As a result, WIT relay stations can also be regarded as special "DEIN UEs". As portrayed in the left cell of Fig. 1.2, both data and energy are simultaneously transferred from DEIN-Station-1 to WIT-Relay-1. The energy harvested by WIT-Relay-1 may be further exploited for forwarding the data packets to the next hop. Since WIT-Relay-2 is beyond the WET range of DEIN-Station-2, it has to request WET from the nearby WET-Station-2 and WET-Station-3. After receiving the data packets from WIT-Relay-1 and gleaning sufficient energy from the WET stations, the data packets are finally forwarded to their destination WIT-UE-1 by WIT-Relay-2.

1.5.3 Heterogeneous Techniques for WIT and WET

Our DEIN architecture has to accommodate both the WET and WIT in the same RF spectral band. Although the WET and WIT both rely on the RF signal, they still have distinctive features, as summarised in Sect. 1.4. Therefore, in order to satisfy the UEs' information and energy requests, the coexistence of WET and WIT in the DEIN results in natural heterogeneity.

In order to guarantee the seamless WIT coverage, different techniques have to be invoked. As exemplified in Fig. 1.2, when the omnidirectional antennas are adopted, the boundary of a DEIN cell is determined by the WIT range of a DEIN station. As a result, the UEs residing within the WIT range of a DEIN station may receive their requested information via a single-hop cellular link. Furthermore, these UEs are also capable of uploading information to their associated DEIN stations. Observe from Fig. 1.2 that WIT-UE-2, DEIN-UE-1 and DEIN-UE-2 all receive their requested information from the downlink WIT of their associated DEIN stations, while WET-UE-2 and DEIN-UE-2 both upload their information to their associated DEIN-Station-2. By contrast, in order to satisfy the information request of a UE sitting beyond the WIT range of a DEIN station, multiple relay stations have to be relied upon for forwarding the information from the DEIN station to the requester or in a reverse direction via the multi-hop transmissions, such as the downlink transmission from DEIN-Station-1 to WIT-UE-1 of Fig. 1.2. In addition, by exploiting the extra energy supplied by the WET stations, a UE beyond the WIT range of a DEIN station is also capable of uploading data to the DEIN station [39], such as the uplink transmission of WET-UE-1 to DEIN-Station-1 in Fig. 1.2, which is powered by WET-Station-1.

If we further look into the wireless charging actions in DEINs, various WET techniques have to be invoked for satisfying diverse charging requirements. As illustrated in Fig. 1.2, a DEIN station's WET range is much shorter than its WIT range, when the omnidirectional antenna is adopted. The reason is that for the successful WET, the energy harvesting circuit of the receiver can only be activated by a high received energy. As a result, the WET is more sensitive to the wireless channel attenuation, which is dominated by the path loss. If a WET UE resides within the WET range of a DEIN station, it may successfully harvest energy from the RF signal emitted by this DEIN station. By contrast, when a WET UE is beyond the WET range, it has to request energy from its nearby WET station. Directional antennas may enable a DEIN station to focus its energy in the main-lobe, which substantially increases the long-range WET efficiency in the direction of the main-lobe. However, the resultant energy loss in the side-lobes may significantly reduce the WET efficiency in other directions. If directional antennas are adopted by the DEIN stations, they may form a narrow energy beam [42] for charging the WET UE beyond the WET range, which is characterised by the omnidirectional antennas. As exemplified in the right DEIN cell of Fig. 1.2, WET-UE-2 is still capable of receiving energy from the dedicated narrow energy beam forming by DEIN-Station-2.

Furthermore, IoT devices will be pervasively deployed in the near future. Our heterogeneous DEINs are also responsible for satisfying both of their communication and energy demands. IoT devices normally are clustered in a specific area in order to jointly carry out their tasks. As a result, for the sake of satisfying the charging requests of the multiple IoT devices, DEIN stations may form wide-angle energy beams for covering the cluster of requesters [43]. This technique may be regarded as energy multicast. Moreover, this wide beam is also capable of transferring information and energy together to the multiple requesters.

References

1. A. Zanella, N. Bui, A. Castellani, L. Vangelista, M. Zorzi, Internet of things for smart cities. IEEE Internet Things J. **1**(1), 22–32 (2014)
2. A. Alrawais, A. Alhothaily, C. Hu, X. Cheng, Fog computing for the internet of things: security and privacy issues. IEEE Internet Comput. **21**(2), 34–42 (2017)
3. B.H. Choi, V.X. Thai, E.S. Lee, J.H. Kim, C.T. Rim, Dipole-coil-based wide-range inductive power transfer systems for wireless sensors. IEEE Trans. Ind. Electron. **63**(5), 3158–3167 (2016)
4. V. Jiwariyavej, T. Imura, Y. Hori, Coupling coefficients estimation of wireless power transfer system via magnetic resonance coupling using information from either side of the system. IEEE J. Emerg. Sel. Top. Power Electron. **3**(1), 191–200 (2015)
5. P.S. Riehl, A. Satyamoorthy, H. Akram, Y.C. Yen, J.C. Yang, B. Juan, C.M. Lee, F.C. Lin, V. Muratov, W. Plumb, P.F. Tustin, Wireless power systems for mobile devices supporting inductive and resonant operating modes. IEEE Trans. Microw. Theory Tech. **63**(3), 780–790 (2015)
6. M. Pinuela, D.C. Yates, S. Lucyszyn, P.D. Mitcheson, Maximizing DC-to-load efficiency for inductive power transfer. IEEE Trans. Power Electron. **28**(5), 2437–2447 (2013)
7. H. Liu, Maximizing efficiency of wireless power transfer with resonant inductive coupling (2011)
8. C. Zheng, H. Ma, J.S. Lai, L. Zhang, Design considerations to reduce gap variation and misalignment effects for the inductive power transfer system. IEEE Trans. Power Electron. **30**(11), 6108–6119 (2015)
9. Z. Yan, Y. Li, C. Zhang, Q. Yang, Influence factors analysis and improvement method on efficiency of wireless power transfer via coupled magnetic resonance. IEEE Trans. Magn. **50**(4), 1–4 (2014)
10. J.M. Miller, O.C. Onar, M. Chinthavali, Primary-side power flow control of wireless power transfer for electric vehicle charging. IEEE J. Emerg. Sel. Top. Power Electron. **3**(1), 147–162 (2015)
11. A. Kurs, A. Karalis, R. Moffatt, J.D. Joannopoulos, P. Fisher, M. Soljačić, Wireless power transfer via strongly coupled magnetic resonances. Science **317**(5834), 83–86 (2007)
12. H. Hwang, J. Moon, B. Lee, C.H. Jeong, S.W. Kim, An analysis of magnetic resonance coupling effects on wireless power transfer by coil inductance and placement. IEEE Trans. Consum. Electron. **60**(2), 203–209 (2014)
13. M.Q. Nguyen, D. Plesa, S. Rao, J.C. Chiao, A multi-input and multi-output wireless energy transfer system, in *2014 IEEE MTT-S International Microwave Symposium (IMS2014)* (2014), pp. 1–3
14. M.Q. Nguyen, Y. Chou, D. Plesa, S. Rao, J.C. Chiao, Multiple-inputs and multiple-outputs wireless power combining and delivering systems. IEEE Trans. Power Electron. **30**(11), 6254–6263 (2015)
15. J.H. Kim, H.Y. Yu, C. Cha, Efficiency enhancement using beam forming array antenna for microwave-based wireless energy transfer, in *2014 IEEE Wireless Power Transfer Conference* (2014), pp. 288–291

16. W.C. Brown, R.H. George, Rectification of microwave power. IEEE Spectr. **1**(10), 92–97 (1964)
17. W.C. Brown, The history of power transmission by radio waves. IEEE Trans. Microw. Theory Tech. **32**(9), 1230–1242 (1984)
18. W.C. Brown, Status of the microwave power transmission components for the solar power satellite. IEEE Trans. Microw. Theory Techn. **29**(12), 1319–1327 (1981)
19. W. Brown, Recent advances in key microwave components that impact the design and deployment of the solar power satellite system, in *1984 Antennas and Propagation Society International Symposium*, vol. 22 (1984), pp. 339–340
20. R.M. Dickinson, Performance of a high-power, 2.388-GHz receiving array in wireless power transmission over 1.54 km, in *1976 IEEE-MTT-S International Microwave Symposium* (1976), pp. 139–141
21. P.S. Yedavalli, T. Riihonen, X. Wang, J.M. Rabaey, Far-field RF wireless power transfer with blind adaptive beamforming for internet of things devices. IEEE Access **5**, 1743–1752 (2017)
22. T. Mitani, S. Kawashima, T. Nishimura, Analysis of voltage doubler behavior of 2.45-GHz voltage doubler-type rectenna. IEEE Trans. Microw. Theory Tech. **65**(4), 1051–1057 (2017)
23. X. Chu, Q. Sun, J. Wang, P. Lu, W.X.X. Xu, Generating a Bessel-Gaussian beam for the application in optical engineering. Scientific Reports **5**(6), 18665 (2015)
24. N. Moraitis, P.N. Vasileiou, C.G. Kakoyiannis, A. Marousis, A.G. Kanatas, P. Constantinou, Radio planning of single-frequency networks for broadcasting digital TV in mixed-terrain regions. IEEE Antennas Propag. Mag. **56**(6), 123–141 (2014)
25. M. Morelli, M. Moretti, A maximum likelihood approach for SSS detection in LTE systems. IEEE Trans. Wirel. Commun. **16**(4), 2423–2433 (2017)
26. L. Sanabria-Russo, J. Barcelo, B. Bellalta, F. Gringoli, A high efficiency MAC protocol for WLANs: providing fairness in dense scenarios. IEEE/ACM Trans. Netw. **25**(1), 492–505 (2017)
27. R. Ford, M. Zhang, M. Mezzavilla, S. Dutta, S. Rangan, M. Zorzi, Achieving ultra-low latency in 5G millimeter wave cellular networks. IEEE Commun. Mag. **55**(3), 196–203 (2017)
28. T. Zhang, L. An, Y. Chen, K.K. Chai, Aggregate interference statistical modeling and user outage analysis of heterogeneous cellular networks, in *2014 IEEE International Conference on Communications (ICC)* (2014), pp. 1260–1265
29. J. Hu, L.L. Yang, L. Hanzo, Distributed multistage cooperative-social-multicast-aided content dissemination in random mobile networks. IEEE Trans. Veh. Technol. **64**(7), 3075–3089 (2015)
30. J. Hu, L.L. Yang, H.V. Poor, L. Hanzo, Bridging the social and wireless networking divide: information dissemination in integrated cellular and opportunistic networks. IEEE Access **3**, 1809–1848 (2015)
31. J. Hu, L.L. Yang, L. Hanzo, Delay analysis of social group multicast-aided content dissemination in cellular system. IEEE Trans. Commun. **64**(4), 1660–1673 (2016)
32. J. Hu, L.L. Yang, L. Hanzo, Energy-efficient cross-layer design of wireless mesh networks for content sharing in online social networks. IEEE Trans. Veh. Technol. **PP**(99), 1–1 (2017)
33. K. Yang, Q. Yu, S. Leng, B. Fan, F. Wu, Data and energy integrated communication networks for wireless big data. IEEE Access **4**, 713–723 (2016)
34. M.D. Renzo, W. Lu, System-Level analysis and optimization of cellular networks with simultaneous wireless information and power transfer: Stochastic geometry modelling. IEEE Trans. Veh. Technol. **66**(3), 2251–2275 (2017)
35. D.H. Chen, Y.C. He, Full-Duplex secure communications in cellular networks with downlink wireless power transfer. IEEE Trans. Commun. **66**(1), 265–277 (2018)
36. J. An, K. Yang, J. Wu, N. Ye, S. Guo, Z. Liao, Achieving sustainable ultra-dense heterogeneous networks for 5g. IEEE Commun. Mag. **55**(12), 84–90 (2017)
37. X. Liu, Z. Li, C. Wang, Secure decode-and-forward relay SWIPT systems with power splitting scheme. IEEE Trans. Veh. Technol. pp. 1–1 (2018)
38. Y. Zhao, D. Wang, J. Hu, K. Yang, H-AP Deployment for joint wireless information and energy transfer in smart cities. IEEE Trans. Veh. Technol. pp. 1–1 (2018)
39. J. Hu, Y. Xue, Q. Yu, K. Yang, A joint time allocation and UE scheduling algorithm for full-duplex wireless powered communication networks. in *2017 IEEE 86th Vehicular Technology Conference (VTC-Fall)* (2017), pp. 1–5

40. K. Lv, J. Hu, Q. Yu, K. Yang, Throughput maximization and fairness assurance in data and energy integrated communication networks. IEEE Internet of Things Journal, **5**(2), 636–644 (2018)
41. Y. Ye, Y. Li, D. Wang, F. Zhou, R.Q. Hu, H. Zhang, Optimal transmission schemes for DF relaying networks using SWIPT. IEEE Trans. Veh. Technol. pp. 1–1 (2018)
42. Y. Zeng, B. Clerckx, R. Zhang, Communications and signals design for wireless power transmission. IEEE Trans. Commun. **65**(5), 2264–2290 (2017)
43. Q. Li, Q. Zhang, J. Qin, Robust tomlinson-harashima precoding with gaussian uncertainties for SWIPT in MIMO broadcast channels. IEEE Trans. Signal Process. **65**(6), 1399–1411 (2017)

Chapter 2
Fundamental of Integrated WET and WIT

Abstract In order to realise integrated wireless energy transfer (WET) and wireless information transfer (WIT), we have to revisit the information theory for finding its performance limits, while redesigning the transceiver architecture in the physical layer for practical implementation. As a result, in this chapter, we impose the energy delivery requirement on the channel output sequence, when maximising the mutual information. The rate-energy tradeoff is studied from the information theoretical perspective for both the discrete-input-discrete-output channel and for the continuous-input-continuous-output channel. Then we provide an overview on the transceiver architecture in the physical layer by considering diverse signal splitters, namely the spatial splitter, the power splitter and the time switcher. The resultant integrated WET and WIT performance is then evaluated for different transceiver architectures.

Keywords Continuous-Input-Continuous-Output-Channel
Discrete-Input-Discrete-Output Channel · Information Theory · Integrated WET and WIT · Multiple-Input-Multiple-Output (MIMO) system · Mutual Information · Power Splitting · Rate-Energy Tradeoff · RF based Wireless Charging · Simultaneous Wireless Information and Power Transfer (SWIPT) Spatial Splitting · Time Switching · Transceiver Architecture · Wireless Energy Transfer (WET) · Wireless Information Transfer (WIT)

In this chapter, we will focus on the fundamental of the physical layer for implementing the integrated wireless energy transfer (WET) and wireless information transfer (WIT) of the point-to-point link. First of all, the information theoretical essence of the integrated WET and WIT will be introduced. Key enabling modules of the generic transceiver architecture for the integrated WET and WIT will also be included. Then, we will cover the architectures of several popular receivers equipped with multiple antennas for simultaneous information and energy reception, namely the ideal receiver, the spatial splitting based receiver, the power splitting based receiver and the time switching based receiver.

J. Hu and K. Yang, *Data and Energy Integrated Communication Networks*, SpringerBriefs in Computer Science, https://doi.org/10.1007/978-981-13-0116-2_2

2.1 Information Theoretical Essence

As previously discussed in Sect. 1.4, WET and WIT entail several conflicting specifications, when they are coordinated in the same radio frequency (RF) spectral band. As a result, theoretical investigations have to be carried out in order to reveal the underlying relationship between the WET and WIT in data and energy integrated communication networks (DEINs), which may provide researchers and engineers with further valuable insights on improving the system-level performance of DEINs. In this section, we will explore the information theoretical essence for DEIN and reveal the natural contradiction between WET and WIT from an information theoretical perspective, which requires further efforts for jointly designing energy and information transfer.

Note that the information theoretical exploration remain valid not only for the integrated WET and WIT operating in the RF spectral band, but for all other integrated data and power transfer scenarios, such as power line communication [1] and power over Ethernet technique [2]. Classic information theoretical channel capacity analysis has been dedicated to maximising the mutual information under the constraint of specific input signals. By contrast, the pioneering work of Gastpar [3] has attempted to maximise the mutual information under the constraint of specific output signals, which aims for controlling the interference imposed by a communicating pair on other peers. This piece of work may provide us with valuable hints for finding the performance limits of integrated data and energy transfer.

2.1.1 Discrete-Input-Discrete-Output Memoryless Channel

We first consider the classic Discrete-Input-Discrete-Output Memoryless Channel (DIDO-MC) of Fig. 2.1. If a DIDO-MC has a single input symbol x and a single output symbol y, the transition probability of this DMC channel may be expressed by the probability function of $p_{\mathcal{Y}|\mathcal{X}}(y|x)$. All the legitimate values of the input symbol x constitute the input codebook $\mathcal{X}$, while all the legitimate values of the output symbol y constitute the output codebook $\mathcal{Y}$. Furthermore, given a specific output symbol y, the energy carried by it can be represented by the non-negative function $g(y)$.

Let us now move on to the n-dimensional random input, which is expressed by the vector $\mathbf{X}^n = (X_1, X_2, \ldots, X_n)$. All the random symbols in the vector $\mathbf{X}^n$

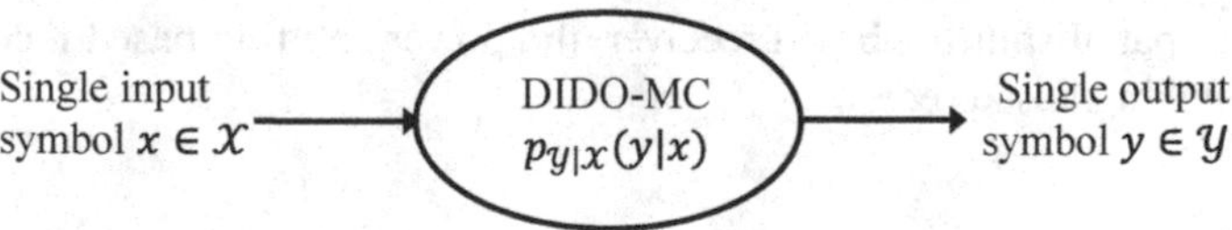

Fig. 2.1 Discrete-input-discrete-output memoryless channel

are independent of one another. If a sample of the n dimensional random input is $\mathbf{x}^n = (x_1, x_2, \ldots, x_n)$, its corresponding occurrence probability can be expressed as $p_{\mathcal{X}^n}(\mathbf{x}^n) = \prod_{i=1}^{n} p_{\mathcal{X}}(x_i)$, where $\mathcal{X}^n$ represents the n-dimensional codebook containing all the possible values of the random output $\mathbf{X}^n$ and $p_{\mathcal{X}}(x_i)$ represents the probability of the symbol x_i being generated. When the input sample $\mathbf{x}^n$ is transmitted by the information source, the corresponding output at the information destination is denoted by the vector $\mathbf{y}^n = (y_1, y_2, \ldots, y_n)$. This sequence of symbols occurs with a probability of $p_{\mathcal{Y}^n}(\mathbf{y}^n) = \prod_{i=1}^{n} p_{\mathcal{Y}}(y_i)$, where $\mathcal{Y}^n$ represents the n-dimensional codebook containing all the possible values of the random output $\mathbf{Y}^n$ and $p_{\dagger}(y_i) = \sum_{x_i \in \mathcal{X}} p_{\mathcal{X}}(x_i) p_{\mathcal{Y}|\mathcal{X}}(y_i|x_i)$ represents the probability of the symbol y_i being received by the information destination. The energy carried by the sequence $\mathbf{y}^n$ can be calculated by the non-negative function $g(\mathbf{y}^n)$. Assuming a random output sequence $\mathbf{Y}^n = (Y_1, Y_2, \ldots, Y_n)$, the average energy carried by this n-dimensional sequence can be formulated as

$$E[g(\mathbf{Y}^n)] = \sum_{\mathbf{y}^n \in \mathcal{Y}^n} g(\mathbf{y}^n) \cdot p_{\mathcal{Y}^n}(\mathbf{y}^n). \tag{2.1}$$

As a result, the WIT performance limit can be formulated as the following optimisation problem:

$$\text{Objective:} \quad \max_{p_{\mathcal{X}^n}(\mathbf{x}^n)} I(\mathbf{X}^n; \mathbf{Y}^n), \tag{2.2}$$

$$\text{Subject to:} \quad \frac{1}{n} \cdot E[g(\mathbf{Y}^n)] \geq \beta, \tag{2.2a}$$

where $I(\mathbf{X}^n; \mathbf{Y}^n)$ is the average mutual information between the n-dimensional input symbol sequence and its output counterpart. Given the transition probabilities $p_{\mathcal{Y}|\mathcal{X}}(\mathbf{Y}|\mathbf{X})$ of the DIDO-MC, the optimisation problem (2.2) aims for finding the optimal n-dimensional information source, which is represented by the probabilities $p_{\mathcal{X}^n}(\mathbf{x}^n)$ of the symbol sequences being generated by this information source. However, this optimisation problem is subject to the condition (2.2a), suggesting that the average energy carried by a single output symbol has to be higher than a threshold β in order to satisfy the basic WET requirement. Substituting the optimal $p_{\mathcal{X}^n}(\mathbf{x}^n)$ into the objective (2.2), we may derive the channel capacity $C_n(\beta)$, when the input is an n-dimensional symbol sequence. Note that $C_n(\beta)$ is a function of the energy constraint β. Furthermore, the normalised channel capacity subject to the energy constraint β can be further formulated as

$$C(\beta) = \sup_n \frac{1}{n} \cdot C_n(\beta), \text{ [bit/symbol]} \tag{2.3}$$

which may be regarded as the rate-energy function. Observe from (2.3) that the rate-energy function is a natural extension of the classic channel capacity concept. This function only depends on a channel's statistical property and on the requirement of the energy harvested, but it does not rely on the information source.

The rate-energy functions of some simple binary channels will now be studied for illustrating the above information theoretical methodology. By adopting the classic ON-OFF-Keying (OOK) as our modulation scheme, for a binary random input, we may assume that the symbol '0' does not carry any energy as $g(0) = 0$, while the symbol '1' carries a single unit of energy as $g(1) = 1$. The probability distribution of the input binary symbols is denoted as $\mathbf{p}_X = \{p_X(0) = q, p_X(1) = \widehat{q} = 1 - q\}$.

As illustrated in Fig. 2.2a, in a noiseless channel, an input binary symbol, either '0' or '1', can be correctly output. Therefore, neither energy loss nor energy gain exist in the noiseless channel. The mutual information of the noiseless channel is $I_N(X;Y) = -q\log_2 q - \widehat{q}\log_2\widehat{q}$. Without any energy constraint, we can maximise $I_N(X;Y)$ by letting $q = 1/2$. Accordingly, the maximum mutual information is $I_{N,\max}(X;Y) = 1$ bit/symbol. Moreover, the average energy carried by the corresponding output symbol is $\beta_{th} = 1/2$ unit. Furthermore, the maximum energy carried by the output symbol at the information destination Y is $\beta_{\max} = 1$ unit, when the source never send the symbol '0', namely $q = 0$. In a nutshell, given a specific energy transfer requirement β, the maximum achievable information rate of the noiseless channel is formulated as [4]

$$C_N(\beta) = \begin{cases} 1, & 0 \le \beta \le \frac{1}{2}, \\ H_2(\beta), & \frac{1}{2} < \beta \le 1, \end{cases} \text{ [bit/symbol]}, \tag{2.4}$$

where $H_2(\beta) = -\beta\log_2\beta - (1-\beta)\log_2(1-\beta)$ is the binary entropy function with respect to β.

As shown in Fig. 2.2b, in a Z channel, the input binary symbol '0' can be correctly output for certain. By contrast, the input binary symbol '1' can be erroneously output as the symbol '0' with a probability ω, while it can be correctly output with a probability $\widehat{\omega} = 1 - \omega$. In this channel, the energy carried by the symbol '1' may be lost during the transmission, due to the channel attenuation. By contrast, since no additional energy is supplied to the system, the input symbol '0' does not have a chance to become the energy carrier symbol '1'. The mutual information of the Z channel is expressed as

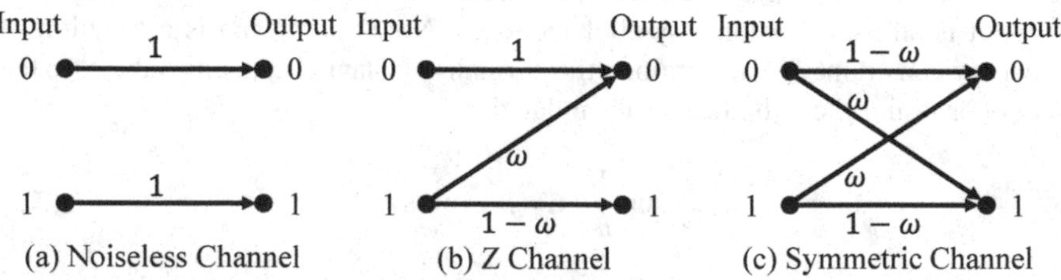

Fig. 2.2 Three typical binary channels

$$I_Z(X;Y) = -(q+\widehat{q}\omega)\log_2(q+\widehat{q}\omega) - \widehat{q}\widehat{\omega}\log_2\widehat{q} + \widehat{q}\omega\log_2\omega. \tag{2.5}$$

Without any energy constraint, we can maximise $I_Z(X;Y)$ by letting

$$q = \frac{1-\omega^{\frac{1}{1-\omega}}}{1+(1-\omega)\omega^{\frac{\omega}{1-\omega}}}. \tag{2.6}$$

Accordingly, the maximum mutual information is $I_{Z,\max} = \log_2(1+\widehat{\omega}\omega^{\frac{\omega}{\widehat{\omega}}})$. Moreover, the average energy carried by the corresponding output symbol is formulated as

$$\beta_{th} = \frac{\omega^{\frac{1}{\widehat{\omega}}}}{1+\widehat{\omega}\omega^{\frac{\omega}{\widehat{\omega}}}} \text{ [unit]}. \tag{2.7}$$

When the source only sends the energy carrier symbol '1', namely $q = 0$, the maximum energy carried by the output symbol at the information destination is $\beta_{\max} = \omega$ unit. In a nutshell, given a specific energy transfer requirement β, the maximum achievable information rate of the Z channel can be formulated as [4]

$$C_Z(\beta) = \begin{cases} \log\left(1-\omega^{\frac{1}{1-\omega}}+\omega^{\frac{\omega}{1-\omega}}\right), & 0 \le \beta \le (1-\omega)\pi^* \\ H_2(\beta) - \frac{\beta}{1-\omega}H_2(\omega), & (1-\omega)\pi^* < \beta \le 1-\omega, \end{cases} \text{ [bit/symbol]}, \tag{2.8}$$

where the variable π^* is given by [4]

$$\pi^* = \frac{\omega^{\frac{\omega}{1-\omega}}}{1+(1-\omega)\omega^{\frac{\omega}{1-\omega}}}. \tag{2.9}$$

As portrayed in Fig. 2.2c, in a symmetric channel, both the input binary symbols '0' and '1' can be erroneously delivered to the output end with a probability of ω, while they can be correctly delivered with a probability of $\widehat{\omega} = 1-\omega$. Apart from the energy loss incurred by the transition from the input symbol '1' to the output symbol '0', the energy of the interference may change the input symbol '0' to the output symbol '1', which results in the energy gain at the information destination. The mutual information of the symmetric channel is expressed as

$$\begin{aligned} I_S(X;Y) =& -(q\widehat{\omega}+\widehat{q}\omega)\log_2(q\widehat{\omega}+\widehat{q}\omega) - (q\omega+\widehat{q}\widehat{\omega})\log_2(q\omega+\widehat{q}\widehat{\omega}) \\ &+ \widehat{\omega}\log_2\widehat{\omega} + \omega\log_2\omega. \end{aligned} \tag{2.10}$$

Without any energy constraint, we can maximise $I_S(X;Y)$ by letting $q = 1/2$. Accordingly, the maximum mutual information is $I_{S,\max} = 1+\widehat{\omega}\log_2\widehat{\omega}+\omega\log_2\omega$. Moreover, the average energy carried by the corresponding output symbol is $\beta_{th} = 1/2$ unit. When the source only sends the energy carrier symbol '1', namely $q = 0$, the maximum energy carried by the output symbol at the information destination is $\beta_{\max} = 1 - \omega$ unit. In a nutshell, given a specific energy transfer requirement β, the maximum achievable information rate of the symmetric channel can be formulated as [4]

$$C_S(\beta) = \begin{cases} \log 2 - H_2(\omega), & 0 \leq \beta \leq \frac{1}{2} \\ H_2(\beta) - H_2(\omega), & \frac{1}{2} < \beta \leq 1 - \omega, \end{cases} \quad \text{[bit/symbol]}, \tag{2.11}$$

where $H_2(\beta)$ and $H_2(\omega)$ are binary entropy functions with respect to β and ω, respectively.

We plot the rate-energy functions of these binary channels in Fig. 2.3. Observe from Fig. 2.3 that given the same energy requirement threshold β, naturally the noiseless channel has the highest information rate, since the information symbols can be transmitted without any error probability over this channel. Furthermore, we also find that given the same energy threshold β, the Z channel may achieve a higher information rate than the symmetric channel, since the Z channel produces fewer errors than the symmetric channel. Furthermore, we denote the optimal distribution of the input symbols for maximising the mutual information without any energy constraint as $\mathbf{p}_X^*$. For these three classic binary channel, when the actual energy requirement β is lower than β_{th}, it has no impact on the achievable information rate, which is also the channel capacity achieved by letting the distribution of the input symbols equal to $\mathbf{p}_X^*$. However, when we have $\beta > \beta_{th}$, we have to change the distribution $\mathbf{p}_X$ of the input symbols for satisfying the required energy constraint, which sabotages the optimality of the original mutual information maximisation. Hence, the achievable information rate is reduced. When we have $\beta > \beta_{\max}$, no realistic scheme is capable of satisfying this demanding energy constraint. Our analysis from the infor-

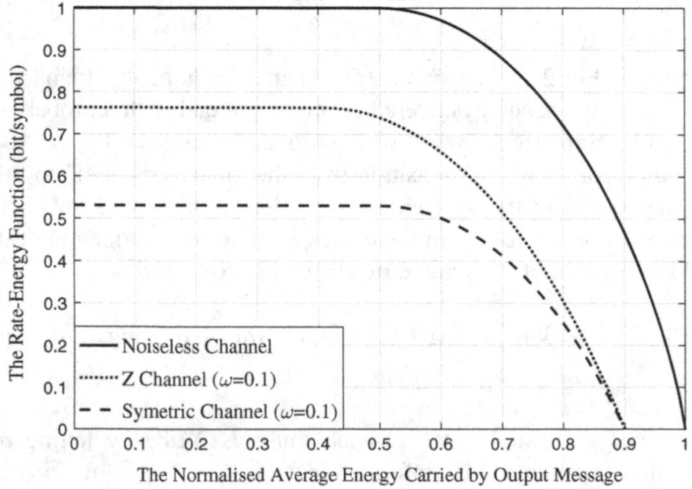

Fig. 2.3 Rate-energy functions for three binary channels

mation theoretical aspect tells us that the careful design of the codewords is essential for the integrated information and energy transfer, which provides the theoretical fundamental for the coding controlled DEIN.

2.1.2 Continuous-Input-Continuous-Output Memoryless Channel

A similar methodology can also be invoked in a continuous-input-continuous-output memoryless channel (CICO-MC), as illustrated in Fig. 2.4. Both the input and the output messages are continuous random variables, which are defined as X and Y, respectively. The transition probability density function of a CICO-MC can be defined as $q_{Y|X}(y|x)$, when the information source randomly generates the continuous input message x and the corresponding continuous output message y is then received by the information destination. All the legitimate values of the random input X constitute the message space $\mathcal{X}$, while the message space $\mathcal{Y}$ contains all the legitimate values of Y. The energy carried by a continuous output message y can be represented by the function $g(y)$.

When the information source transmits a message $x \in \mathcal{X}$, the average energy received by the information destination can be formulated as

$$\rho(x) = \int_{\mathcal{Y}} q_{Y|X}(y|x) \cdot g(y)dy. \tag{2.12}$$

After taking the expectation operation on $\rho(x)$, the average energy received by the information destination can be expressed as

$$\begin{aligned} E[\rho(x)] &= \int_{\mathcal{X}} \rho(x)dF_X(x) = \int_{\mathcal{X}} dF_X(x) \int_{\mathcal{Y}} q_{Y|X}(y|x) \cdot g(y)dy \\ &= \int_{\mathcal{X}} \int_{\mathcal{Y}} q_{Y|X}(y|x) \cdot g(y) \cdot dF_X(x)dy = E[g(y)], \end{aligned} \tag{2.13}$$

where $F_X(x)$ is the cumulative distribution function (CDF) of the random input message X.

Similar to its DIDO-MCs counterpart, the rate-energy function of CICO-MCs can also be expressed by the following optimisation problem:

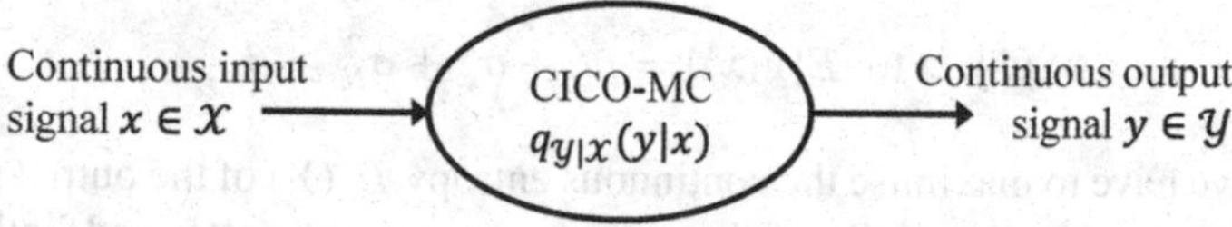

Fig. 2.4 Continuous-input-continuous-output memoryless channel

$$\text{Objective: } \max_{F_X(x)} I(X;Y), \tag{2.14}$$

$$\text{Subject to: } E[\rho(x)] \geq \beta, \tag{2.14a}$$

where the continuous random variable X is the input message, the continuous random variable Y is the corresponding output message, $I(X;Y)$ is their average mutual information and β is the minimum energy requirement of the information destination. The optimisation problem (2.14) aims for finding a test information source X^*, whose CDF $F_{X^*}(x)$ is capable of maximising the average mutual information $I(X;Y)$, while fulfilling the minimum energy requirement β. Upon substituting the optimal CDF $F_{X^*}(x)$ into (2.14), we may obtain the rate-energy function of the CICO-MC. Assuming that the input message falls into the region $[-\alpha, \alpha]$, the rate-energy function can be defined as $C(\alpha, \beta)$.

Let us now study the case of Additive White Gaussian Noise (AWGN) channel as an example of the CICO-MC. In the classic AWGN channel, the amplitude N of the noise follows a Gaussian distribution having a mean of zero and a variance of σ_N^2. When the information destination receives the message y, the energy carried by this message is $g(y) = y^2$. When the input message x is transmitted by the information source, the average energy received by the destination may be expressed as

$$g(x) = \int_{-\infty}^{+\infty} \frac{y^2}{\sqrt{2\pi\sigma_N^2}} e^{-\frac{(y-x)^2}{2\sigma_N^2}} dy = x^2 + \sigma_N^2. \tag{2.15}$$

Observe from (2.15) that the energy harvested by the information destination is constituted by two parts, namely the energy carried by the input message x, which is expressed as x^2, and the channel's noise energy, which is expressed as σ_N^2. Upone averaging $g(x)$ of (2.15) over all legitimate values of the input message x, we arrive at the average energy that can be harvested by the information destination in an AWGN channel, which can be expressed as

$$E[g(x)] = E[x^2 + \sigma_N^2] = \mu_x^2 + \sigma_x^2 + \sigma_N^2, \tag{2.16}$$

where μ_x is the mean of the random input message X and σ_x^2 is its variance.

By considering the AWGN channel, the optimisation problem (2.14) can be expressed as

$$\text{Objective: } \max_{F_X(x)} H_c(Y) - H_c(N) = H_c(Y) - \frac{1}{2}\log 2\pi e\sigma_N^2 \tag{2.17}$$

$$\text{Subject to: } E[g(x)] = \mu_x^2 + \sigma_x^2 + \sigma_N^2 \geq \beta. \tag{2.17a}$$

Naturally, we have to maximise the continuous entropy $H_c(Y)$ of the output message Y. Since the average power P_Y of the output message is constrained both by the average power of the input message as well as by the power of the noise, $H_c(Y)$ can be maximised, when Y is a Gaussian distributed random variable having a zero mean and a variance of $\sigma_y^2 = P_Y$. Since the noise N is also Gaussian distributed, the input message X has to be a Gaussian distributed random variable having a zero mean as well. Furthermore, according to the constraint (2.17a), the variance of X should follow $\sigma_x^2 \geq (\beta - \sigma_N^2)$. Since the variance of the output message Y is $\sigma_y^2 = \sigma_x^2 + \sigma_N^2$, the rate-energy function of the AWGN channel can be expressed as

$$\begin{aligned} C_{AWGN}(\beta) &= H_{c,\max}(Y) - \frac{1}{2}\log 2\pi e\sigma_N^2 \\ &= \frac{1}{2}\log 2\pi e\sigma_y^2 - \frac{1}{2}\log 2\pi e\sigma_N^2 \\ &= \frac{1}{2}\log\frac{\sigma_x^2 + \sigma_N^2}{\sigma_N^2} = \frac{1}{2}\log\frac{\beta}{\sigma_N^2}. \end{aligned} \tag{2.18}$$

2.2 Transceiver Architecture of DEIN Devices

In order to practically implement the integrated WET and WIT function in the physical layer, the transceiver architecture should be redesigned in order to satisfy both the WET and WIT requirements. A typical transceiver architecture for the integrated WET and WIT is illustrated in Fig. 2.5. Specifically, the transmitter of Fig. 2.5 consists of the following modules:

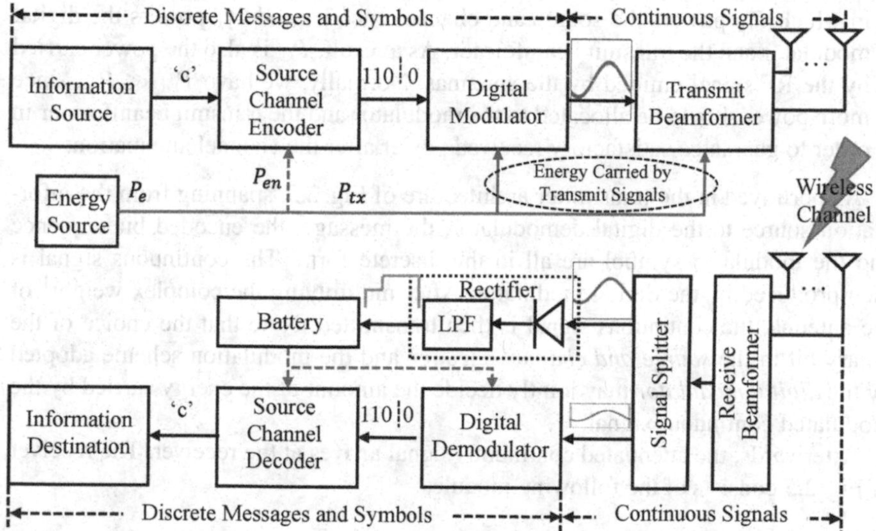

Fig. 2.5 Transceiver architecture of DEIN devices

- *Information Source*, which generates discrete messages to be transmitted. For instance, the information source X of Fig. 2.5 may generate a message from the set {a,b,c,d} by obeying the probability distribution of $\{p(\text{a}) = \frac{1}{2}, p(\text{b}) = \frac{1}{4}, p(\text{c}) = \frac{1}{8}, p(\text{d}) = \frac{1}{8}\}$.
- *Source and Channel Encoder*, which encodes the message into binary bits and adds redundant bits for increasing the transmission reliability. In the example of Fig. 2.5, the message 'c' is encoded by the classic Huffman coding scheme into the bit sequence '110', while the redundant bit '0' is also added for the parity check. Hence, the output of the source channel encoder is a bit sequence '1100'.
- *Digital Modulator*, which modulates the bit sequence onto the carrier signals for the transmission. In the example of Fig. 2.5, when the 16-quadrature-amplitude-modulation (16-QAM) is adopted by the modulator, the bit sequence '1100' is modulated as a complex symbol $(-\frac{1}{2}d, -\frac{1}{2}d)$, where d represents the minimum distance between symbol points in the constellation. As a result, the corresponding pass-band signal is expressed as $x(t) = -\frac{1}{2}d\cos(2\pi f_c t + \phi) - \frac{1}{2}d\sin(2\pi f_c t + \phi)$, where f_c and ϕ represent the frequency and the phase of the carrier, respectively. The alphabet of any modulation scheme only consists of finite symbols.
- *Transmit Beamformer*, which is designed for counteracting the wireless channel degradation in the multiple-input-multiple-output (MIMO) system. Based on the knowledge of the channel states, represented by the channel's amplitude attenuation and phase change on the RF signal transmitted, the transmit beamformer adjusts the complex weight $v_{tx}e^{j\theta_{tx}}$ of each transmit antenna, where v_{tx} and θ_{tx} are the amplitude weight and the phase of the antenna, respectively, in order to minimise the adverse effect of the wireless channel degradation as low as possible.
- *Energy Source*, which steadily supplies energy to the other functional modules of the transmitter. Total power provided by the energy source is $P_e = P_{en} + P_{tx}$, in which P_{en} powers the source and channel encoder and P_{tx} powers the digital modulator and the transmit beamformer. As a result, P_{tx} is also the power carried by the RF signal emitted by the antennas. Normally, we have $P_{en} \ll P_{tx}$, since more power should be allocated to the modulator and the transmit beamformer in order to guarantee satisfactory received power after the channel attenuation.

As portrayed in the transmitter architecture of Fig. 2.5, spanning from the information source to the digital demodulator, the message, the encoded bit sequence and the modulated symbol are all in the discrete form. The continuous signal is then produced by the digital modulator. After multiplying the complex weight[1] of the antenna, the continuous signal is then transmitted. Note that the choice of the binary bit in the *source and channel encoder* and the modulation scheme adopted by the *digital modulator* may jointly decide the amount of the energy carried by the modulated continuous signal.

Afterwards, the attenuated continuous signal arrives at the receiver. The receiver of Fig. 2.5 consists of the following modules:

[1]The complex weight include both the amplitude weight and the phase.

- *Receive Beamformer*, which is exploited for compensating for the channel attenuation in the MIMO system. Upon adjusting the complex weight $v_{rx}e^{j\theta_{rx}}$ of each receive antenna, the received signal can be optimally combined in order to increase the achievable throughput and reduce the symbol-error-ratio (SER).
- *Signal Splitter*, which is invoked for splitting the received RF signal into two portions. One portion flows into the energy harvesting circuit in order to charge the battery of the receiver, while the other flows into the information reception circuit in order to recover the contaminated information bits. The signal splitter can operate in the power domain, which yields the power splitter. The signal splitter can also operate in the time domain, which yields the time switcher. Furthermore, we may also design the signal splitter in the spatial domain of the MIMO system, in which the antennas are partitioned into two groups, one for the dedicated energy harvesting and the other for the dedicated information reception.

After the *signal splitter*, one portion of the received continuous signal flows into the information decoding component, which consists of the following modules:

- *Digital Demodulator*, which recovers the information content from the received continuous signal. The coherent and non-coherent detections can be exploited in this module for different applications. Specifically, the coherent detection requires exact knowledge of the channel state for the demodulation, while the non-coherent detection does not. As a result, the coherent detection is capable of achieving better error correction performance than its non-coherent counterpart. However, the low-complexity non-coherent detection is suitable for the communication of the miniature-sized IoT devices.
- *Source and Channel Decoder*, which decode the information bit sequence and recover the original message.
- *Information Destination*, which finally receive the decoded message.

The other portion of the received continuous signal flows into the energy harvesting component, which consists of the following modules:

- *Rectifier*, which rectifies the continuous RF signal into the direct current (DC) signal. As exemplified in Fig. 2.5, a simple rectifier is comprised of a diode and a low-pass-filter (LPF). The output DC signal of the rectifier can be expressed as $i_{DC}(t) = \sum_{n=0}^{+\infty} k_{2n} y_e^{2n}(t)$, where the coefficients $\{k_{2n} | n = 0, 1, \ldots, +\infty\}$ are determined by the electrical property of the diode and $y_e(t)$ is the portion of the continuous received signal flowing into the energy harvesting components. Note that the power P_e of the signal $y_e(t)$ is very small due to the channel attenuation. Therefore, we may reasonably ignore the terms having $y_e(t)$ of high order. By further letting $k_0 = 0$, we may have a linear rectifier $i_{DC}(t) \approx k_2 y_e^2(t)$ having a conversion efficiency of k_2.
- *Battery*, which stores the energy harvested and powers the other modules of the receiver. Normally, the energy gleaned by the battery is linearly proportional to the DC signal $i_{DC}(t)$.

2.3 Signal Splitter Based Receiver Architecture

As portrayed in Fig. 2.5, a decent signal splitter is required at the receiver for realising the simultaneous information and energy reception. In this section, we will firstly introduce the ideal receiver for the integrated information and energy reception, while elaborating on the practical signal splitters in spatial, power and time domain.

2.3.1 *Ideal Receiver Architecture*

Varshny and Grover [4, 5] studied an ideal receiver architecture for integrated WET and WIT. This ideal receiver is capable of simultaneously harvesting energy and decoding information from the same received RF signals, as detailed in Sect. 2.1 from an information theoretical perspective. In this receiver, the information decoder can be regarded as an "observer", which only "observe" the amplitude and phase patterns of the received RF signal in order to recover the original information. As a result, there is no energy loss during the information decoding stage. The energy carried by the received RF signal can thus be completely harvested for charging the batteries or driving the electronic loads. However, this ideal receiver cannot be practically realised. The reason is that the energy carried by the received RF signal may be totally lost at the information decoder, which cannot be further reused for the energy harvesting. Nonetheless, the above ideal receiver architecture is capable of shedding light on the performance limits of the integrated WET and WIT. For example, by invoking this ideal receiver architecture, Zhou and Xiang [6, 7] found the upper bound of the rate-energy region, which characterises the fundamental tradeoff between the WET and WIT.

2.3.2 *Spatial Splitting Based Receiver*

In the spatial splitting (SS) based receiver architecture, as illustrated in Fig. 2.6, the signal is split in the spatial domain by allocating different number of antennas at the receiver end for energy and information reception, respectively. Let us assume that N_r antennas in total are used at the receiver. Given the splitting factor ρ, we may choose a set $\mathcal{A}_{r,e} = \{A_{r,i} | 1 \leq i \leq \lfloor \rho N_r \rfloor\}$ of antennas for the dedicated energy harvesting, while the rest are adopted for information decoding, which form another set $\mathcal{A}_{r,d} = \{A_{r,j} | 1 \leq j \leq N_r - \lfloor \rho N_r \rfloor\}$. Furthermore, if we have N_t transmit antennas on the transmitter side, which are represented $\mathcal{A}_t = \{A_{t,n} | 1 \leq n \leq N_t\}$, then we have $N_t \times \lfloor \rho N_r \rfloor$ MIMO channels for the WET and $N_t \times (N_r - \lfloor \rho N_r \rfloor)$ MIMO channels for the WIT. Let us assume that the signal $y_{A_{r,i}}$ is received by the receive antenna $A_{r,i}$:

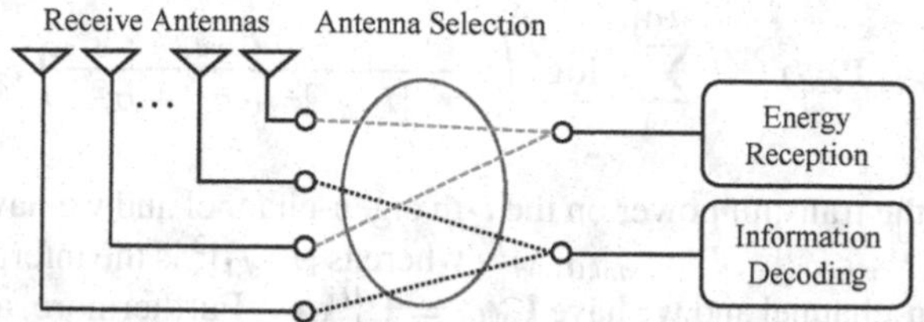

Fig. 2.6 Spatial splitting based receiver architecture. The combined signal received by the antenna group $\mathcal{A}_{r,e}$ is exploited for energy harvesting, while that received by the antenna group $\mathcal{A}_{r,d}$ is exploited for information decoding

$$y_{A_{r,i}} = \sum_{A_{t,j} \in \mathcal{A}_t} h_{A_{t,j},A_{r,i}} x_{A_{t,j}} + I_{A_{r,i}} + z_{0,A_{r,i}} + z_{cov,A_{r,i}}, \tag{2.19}$$

where $x_{A_{t,j}}$ is the modulated symbol transmitted by the antenna $A_{t,j}$, $h_{A_{t,j},A_{r,i}}$ is the channel coefficient between the transmit antenna $A_{t,j}$ and the receive antenna $A_{r,i}$, $I_{A_{r,i}}$ is the total interference imposed on the receive antenna $A_{r,i}$, $z_{0,A_{r,i}}$ is the AWGN having a variance (power) of σ_0^2 imposed on the receive antenna $A_{r,i}$ and finally, $z_{cov,A_{r,i}}$ is the additional AWGN introduced by the passband to baseband converter having a power of σ_{cov}^2. Furthermore, if the transmit power of the transmit antenna $A_{t,j}$ is denoted as $P_{t,A_{t,j}} = |x_{A_{t,j}}|^2$, we can collect the transmit power of all the transmit antennas into a column vector $\mathbf{P}_t = (P_{t,A_{t,1}}, \ldots, P_{t,A_{t,N_t}})^T$.

As for the energy harvesting, both the interference $I_{A_{r,i}}$ and the noise $z_{A_{r,i}}$ can be regarded as useful energy sources. Therefore, the total RF power collected by the receiver should be expressed as

$$P_e^{SS}(\mathcal{A}_{r,e}, \mathbf{P}_t) = \eta_e \sum_{A_{r,i} \in \mathcal{A}_{r,e}} \left(\left| \sum_{A_{t,j} \in \mathcal{A}_t} h_{A_{t,j},A_{r,i}} x_{A_{t,j}} \right|^2 + |I_{A_{r,i}}|^2 + \sigma_{0,A_{r,i}}^2 \right), \tag{2.20}$$

where η_e is the constant efficiency of the receiver's energy harvester. Since $|x_{A_{t,j}}|^2 = P_{t,A_{t,j}}$, $P_{e,total}^{SS}(\mathcal{A}_{r,e}, \mathbf{P}_t)$ is a function with respect to the transmit power vector $\mathbf{P}_t$. This is equivalent to the equal gain combining (EGC) technique widely adopted for the receiver diversity in communication engineering.

By taking the singular-value-decomposition (SVD) on the $N_t \times (N_r - \lfloor \rho N_r \rfloor)$-element MIMO channel $\mathbf{H}_I$, we have $\mathbf{H}_I = \mathbf{U}_I \mathbf{\Lambda}_I \mathbf{V}_I$, where $\mathbf{\Lambda}_I$ is the diagonal matrix constituted by the singular values of $\mathbf{H}_I$. The i-th diagonal element is represented by λ_i. Therefore, the original MIMO channel $\mathbf{H}_I$ is decomposed into $rank(\mathbf{H}_I)$ independent and parallel eigen-channels. Assuming that the transmitter has perfect channel state information (CSI), the achievable information rate can be expressed as

$$R^{SS}(\mathcal{A}_{r,I}, \mathbf{P}_{svd}) = \sum_{i=1}^{rank(\mathbf{H}_I)} \log_2\left(1 + \frac{\lambda_i^2 P_{svd,i}}{|I_{svd,i}|^2 + \sigma_0^2 + \sigma_{cov}^2}\right), \tag{2.21}$$

where $P_{svd,i}$ is the transmit power on the i-th eigen-channel and we have the column vector $\mathbf{P}_{svd} = (P_{svd,i}, \ldots, P_{svd,rank(\mathbf{H}_I)})^T$, whereas $|I_{svd,i}|^2$ is the interference power on the i-th eigen-channel and we have $\mathbf{I}_{svd,i} = \mathbf{U}_I^H \mathbf{I}_{\mathcal{A}_{r,I}}$. Furthermore, after the SVD, we have $\mathbf{P}_t = |\mathbf{V}_I^H|^2 \mathbf{P}_{svd}$. Again, the additional noise power σ_{cov}^2 on each eigen-channel is introduced by the pass-band to baseband converter.

The spatial splitter has become a popular methodology for realising the simultaneuous information and energy reception. In [8], by invoking an antenna selection scheme and by aligning interference at the reduced signal subspace, Koo et al. split the received signal into two orthogonal spaces in order to achieve simultaneous information and energy transfer. Zhao et al. [9] formulated the joint optimisation problem of antenna selection and transmit covariance matrix for the sake of achieving the maximum attainable information rate, subject to the specific energy harvesting constraint encountered.

Apart from the antenna selection, dual-polarised antennas can also be invoked for splitting RF signals in the spatial domain, which are capable of providing polarisation diversity, whilst alleviating the antennas' space limitation. More explicitly, two orthogonal polarisations can be created, which potentially halves the number of antennas [10]. With the aid of dual-polarised antennas adopted for integrated WET and WIT, the data signals and energy signals can thus be transmitted separately. Furthermore, both the data and energy signals can also be received independently with the aid of vertical and horizontal polarisation. Hence, the dual-polarised antenna aided integrated WET and WIT is worth further investigation due to its low-complexity and promising performance.

2.3.3 Power Splitting Based Receiver

In the power splitting (PS) based receiver architecture, the received RF signal can be split in the power domain for supporting the simultaneous information and energy reception, as portrayed in Fig. 2.7.

Let us still consider a generic MIMO channel for the case of the power splitting based receiver, where we have N_t transmit antennas and N_r receive antennas. The MIMO channel can be represented by the $(N_r \times N_t)$-element matrix $\mathbf{H}$. We assume that the symbols transmitted by the transmitter are hosted by a $N_t \times 1$ column vector $\mathbf{x}$. Moreover, the channel coefficients in $\mathbf{H}$ are assumed to remain unchanged during a single symbol. Then we can derive the following equations for characterising the symbols received by the MIMO receiver:

$$\mathbf{y} = \mathbf{H}\mathbf{x} + \mathbf{I} + \mathbf{z}_0 + \mathbf{z}_{cov}, \tag{2.22}$$

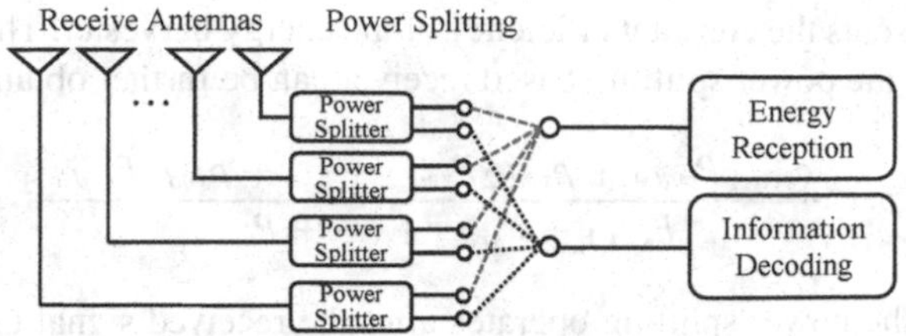

Fig. 2.7 Power splitting based receiver architecture. The received signal is split in the power domain. The energy signal is exploited for energy harvesting, while the data signal is exploited for information decoding

where $\mathbf{I}$ is a $(N_r \times 1)$-element complex column vector representing the complex-valued interference at the receiver, $\mathbf{z}_0$ is a $(N_r \times 1)$-element complex column vector representing the complex AWGN imposed on the receive antennas with a power of σ_0^2 and $\mathbf{z}_{cov}$ is also a $N_r \times 1$ complex column vector representing the complex AWGN having a power of σ_{cov}^2 introduced by the pass-band to baseband converter. Similar to the spatial splitting based receiver of Sect. 2.3.2, we have $\mathbf{H} = \mathbf{U}\boldsymbol{\Lambda}\mathbf{V}$ by taking the SVD on the channel matrix $\mathbf{H}$, where the diagonal matrix $\boldsymbol{\Lambda}$ holds all the singular values of $\mathbf{H}$. These singular values are denoted by $\{\lambda_i | 1 \leq i \leq rank(\mathbf{H})\}$. Upon substituting $\mathbf{y}_{svd} = \mathbf{U}^H\mathbf{y}$ and $\mathbf{x}_{svd} = \mathbf{V}\mathbf{x}$ as well as $\mathbf{I}_s vd = \mathbf{U}^H\mathbf{I}, \mathbf{z}_{0,svd} = \mathbf{U}^H\mathbf{z}_0$ and $\mathbf{z}_{cov,svd} = \mathbf{U}^H\mathbf{z}_{cov}$ into Eq. (2.22), we then have

$$\mathbf{y}_{svd} = \boldsymbol{\Lambda}\mathbf{x}_{svd} + \mathbf{z}_{0,svd} + \mathbf{z}_{cov,svd}, \tag{2.23}$$

where $\mathbf{U}$ and $\mathbf{V}$ are both unitary matrices and hence multiplying the symbol vectors $\mathbf{y}$, $\mathbf{x}$ and $\mathbf{z}_0$ as well as $\mathbf{z}_{cov}$ by them does not alter their statistical properties. Furthermore, the SVD virtually decompose the MIMO channel into several independent eigen-channels. As a result, the capacity of the MIMO channel $\mathbf{H}$ can be further formulated as

$$R^{PS}(\mathbf{P}_{svd}, \boldsymbol{\rho}_{svd}) = \sum_{i=1}^{rank(\mathbf{H})} \log_2\left(1 + \frac{(1-\rho_{svd,i})\lambda_i^2 P_{svd,i}}{(1-\rho_{svd,i})(|I_{svd,i}|^2 + \sigma_0^2) + \sigma_{cov}^2}\right), \tag{2.24}$$

where $\mathbf{P}_{svd} = (P_{svd,1}, \ldots, P_{svd,rank(\mathbf{H})})^T$ and $\boldsymbol{\rho}_{svd} = (\rho_{svd,1}, \ldots, \rho_{svd,rank(\mathbf{H})})^T$ contain the transmit powers and the power splitting factors for the corresponding eigen-channels, respectively. Specifically, in the i-th eigen-channel, the portion $\rho_{svd,i}$ of the received signal is exploited for energy harvesting, while the portion $(1 - \rho_{svd,i})$ of the received signal is exploited for information decoding. The eigenvalue λ_i can be regarded as the coefficient of the i-th eigen-channel. Furthermore, the total energy received from the RF signal can then be derived as

$$P_E^{PS}(\mathbf{P}_{svd}, \boldsymbol{\rho}_{svd}) = \eta_E \sum_{i=1}^{rank(\mathbf{H})} \rho_{svd,i}(\lambda_i^2 P_{svd,i} + |I_{svd,i}|^2 + \sigma_0^2), \tag{2.25}$$

where η_E represents the constant efficiency of the energy harvester. The signal splitting factor ρ of the power splitting based receiver can be further obtained as

$$\rho = \frac{\rho_{svd,1} P_{svd,1} + \rho_{svd,2} P_{svd,2} + \cdots + \rho_{svd,i} P_{svd,i}}{P_{svd,1} + P_{svd,2} + \cdots + P_{svd,i}}. \tag{2.26}$$

In this model, the power splitting operates after the received signal of the multiple antennas is multiplied by a unitary matrix $\mathbf{U}^H$. This operation is equivalent to splitting the received signals of each eigen-channel in the power domain.

There is another pair of power splitting based receiver architectures, namely the independent power splitting [11] and the uniform power splitting [12]. For the independent power splitting, the received signal of each antenna is directly split into two portions for energy harvesting and information decoding, respectively, without any signal processing operation. Furthermore, the antenna selection based spatial splitting receiver architecture of Sect. 2.3.1 can also be regarded as a special independent power splitting, when the power splitting factor at each antenna is either 0 or 1. For the uniform power splitting, the received signals of multiple antennas are firstly combined and the aggregate signal is then split in the power domain for dedicated energy harvesting and information decoding.

The power splitting based receiver architecture has been widely adopted [11, 13–16] for integrating the WET and the WIT in the downlink. Liu et al. [13] have proposed a dynamic power splitting based receiver architecture and they have demonstrated that this receiver outperforms its fixed splitting-based counterpart. Furthermore, Zhang et al. [14] have analysed the attainable rate-energy region of the power splitting based receiver and compared it to its spatial splitting based counterpart. Li et al. [11] have considered the interference-contaminated single-input-multiple-output (SIMO) system for integrated WET and WIT. Their results have validated that the independent power splitting outperforms its uniform counterpart. Moreover, in full-duplex relay networks, the power splitting based receiver architecture is invoked by relays in order to simultaneously receive both energy and information from the sources [15]. The energy received may be depleted by the relays for altruistically forwarding data to their destination. Physical layer security is discussed in the context of a multiple-input-single-output (MISO) system in [16], where power splitting based receiver architecture is exploited for simultaneously receiving both the information and energy.

2.3.4 *Time Switching Based Receiver*

In the time switching (TS) based receiver, the signal received by the multiple antennas is split in the time domain by a time switcher for the independent energy harvesting and the information decoding, as portrayed in Fig. 2.8.

In contrast to the complex power splitter of Fig. 2.7, we only have to partition a typical transmission frame into two parts, namely the WET frame and the WIT frame. The RF signal received during the WET frame is delivered to the energy harvesting

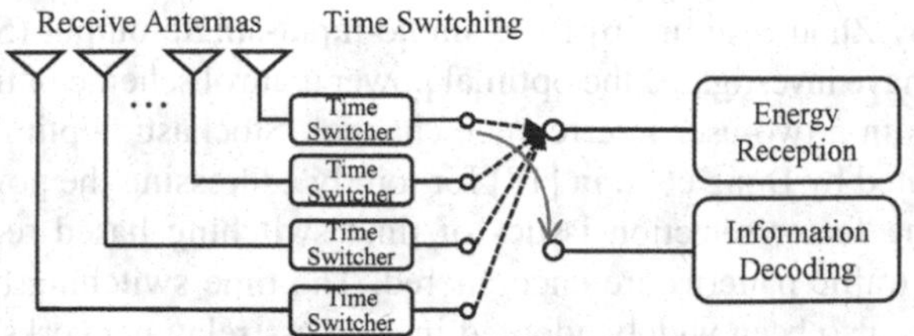

Fig. 2.8 Time switching based receiver architecture. The received signal is split in the time domain. The receiver firstly opts to exploit the received signal for information decoding. Then the receiver switches to the energy harvesting mode and the remaining signal is exploited for charging the energy storage unit

components, as exemplified by Fig. 2.8, in order to convert the energy of the RF signal to DC, while the RF signal received during the WIT frame is delivered to the information decoding components for the sake of recovering the original information.

In the time switching scheme, we consider the same received MIMO signal model as expressed in Eqs. (2.22) and (2.23). As a benefit of its flexibility, we can approximately tune the power splitter for each eigen-channel in order to maximise the information transmission rate or the amount of energy received. By contrast, having diverse time switching factors for each eigen-channel is not quite realistic, since this operation may result in the time domain misalignment of the RF signals received by the independent eigen-channels. Hence, additional synchronisation techniques are required, if we want to assign a specific time switching factor for each eigen-channel, which may significantly complicate the structural design of the transmission frame and the actual implementation of the diverse tunable time switcher.

As a result, we consider an identical time switching factor for all the independent eigen-channels, which indicates that only a single time switcher is required. Considering a time switching factor of ρ, the amount of power that can be harvested from a portion of the received RF signal to the DC is expressed as

$$P_e^{TS}(\mathbf{P}_{svd}, \rho) = \eta_e \cdot \rho \cdot \sum_{i=1}^{rank(\mathbf{H})} (\lambda_i^2 P_{svd,i} + |I_{svd,i}|^2 + \sigma_0^2). \tag{2.27}$$

Furthermore, the rest of the received RF signal is exploited for the information decoding. As a result, the attainable information transmission rate can be derived as

$$R^{TS}(\mathbf{P}_{svd}, \rho) = (1-\rho) \sum_{i=1}^{rank(\mathbf{H})} \log_2 \left(1 + \frac{\lambda_i^2 P_{svd,i}}{|I_{svd,i}|^2 + \sigma_0^2 + \sigma_{cov}^2}\right), \tag{2.28}$$

where $\mathbf{P}_{svd} = (P_{svd,1}, \ldots, P_{svd,rank(\mathbf{H})})^T$.

Due to its easy implementation, the time switching based receiver architecture has attracted abundant of interests in the context of the integrated WET and WIT [6, 17–20]. The rate-energy region of the time switching based receiver architecture has been characterised by Zhou et al. in [6] for a single-input-single-output (SISO) system. Zhu et al. [17] have investigated the optimal power control scheme of time switching based receivers in a two-user interference channel. Stochastic optimisation theory has been exploited by Dong et al. in [18] for jointly addressing the power allocation and the splitting factor selection issues of time switching based receivers, when heterogeneous traffic patterns are encountered. The time switching based receiver architecture has also been widely adopted in wireless relay networks [19, 20]. For example, a relay selection scheme was jointly designed with the selection of time switching splitting factors by Atapattu et al. in [19] for the sake of optimising outage probabilities and end-to-end data throughput. Furthermore, sub-carriers allocation as well as selection of time switching factors have been jointly optimised for multiple transmitter-receiver pairs [20].

2.3.5 Rate-Energy Region

For the signal splitter based receivers of Figs. 2.6, 2.7 and 2.8, a tradeoff has to be struck between the information rate and the amount of energy harvested. As the signal splitting factor ρ increases, more RF-signal energy is exploited for the energy harvesting, which inevitably diminishes the attainable information rate. By contrast, as the signal splitting factor ρ reduces, more RF-signal energy is used for the information decoding, which degrades the energy harvesting performance.

The achievable rate-energy (R-E) region is capable of characterising the performance of the integrated WET and WIT. We can now obtain the rate-energy region of the power splitting based receiver with the aid of Eqs. (2.24) and (2.25) for the SISO channel considered as

$$\mathcal{C}_{R-E}^{PS}(P_t) \triangleq \bigcup_{\rho} \Big\{ (R^{PS}, P_E^{PS}) : P_E^{PS} \leq \eta_E \rho (|h|^2 P_t + |I|^2 + \sigma_0^2), \quad R^{PS} \leq \log_2 \left(1 + \frac{(1-\rho)|h|^2 P_t}{(1-\rho)(|I|^2 + \sigma_0^2) + \sigma_{cov}^2} \right) \Big\}. \tag{2.29}$$

Similarly, with the aid of Eqs. (2.28) and (2.27), the rate-energy region of the time switching based receiver is formulated for the SISO channel as

$$\mathcal{C}_{R-E}^{TS}(P_t) \triangleq \bigcup_{\rho} \Big\{ (R^{TS}, P_E^{TS}) : P_E^{PS} \leq \eta_E \rho (|h|^2 P_t + |I|^2 + \sigma_0^2), \quad R^{TS} \leq (1-\rho) \log_2 \left(1 + \frac{|h|^2 P_t}{|I|^2 + \sigma_0^2 + \sigma_{cov}^2} \right) \Big\} \tag{2.30}$$

Figure 2.9 clearly characterise the rate-energy tradeoff for both the power splitting based receiver of Fig. 2.7 and its time switching based counterpart of Fig. 2.8. The

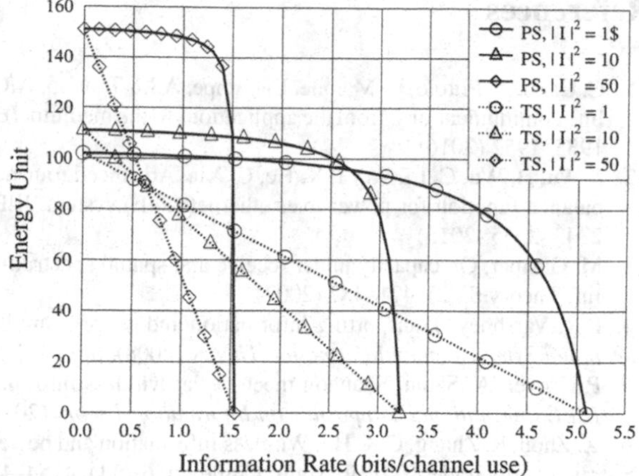

Fig. 2.9 Rate-energy tradeoff for the time switching based receiver and the power splitting based receiver in the SISO channel having $P_t = 100$, $|h|^2 = 1$, $\sigma_0^2 = 1$, $\sigma_{cov}^2 = 1$, $\eta_E = 1$ and $|I|^2 = \{1, 10, 50\}$

parameters in Fig. 2.9 are normalised by the AWGN power σ_0^2 at the receive antenna. We can observe from Fig. 2.9 that the interference does not always impose adverse effects on the integrated WET and WIT system. For example, for the power splitting based receiver of Fig. 2.7, if the desired information rate is as low as 1 bit/channel use, we can gain more energy when we have more interference at the receiver. Similarly for the time switching based receiver of Fig. 2.8, the same trend can be observed if the desired information rate is as low as 0.5 bit/channel user. This observation provides us some hints on the implementation issues of the integrated WET and WIT. We can enhance the transmit power for the low-rate services, such as paging or for the control signalling exchange in cellular communications, in order to transfer more energy to the receiver. In this scenario, the energy reception benefits from the inter-cell interference or other types of interference without violating the quality of the low-rate service.

2.4 Summary

In this chapter, we firstly overviewed the information theoretical fundamental of the integrated data and energy transfer in the discrete-input-discrete-output memoryless channel and the continuous-input-continuous-output channel. Key enabling modules of the generic transceiver architecture for the integrated WET and WIT were introduced. Furthermore, we focused our attention on the signal splitter based receiver architecture equipped with multiple antennas, which includes the spatial splitting based receiver, power splitting based receiver and the time switching based receiver. Finally, the rate-energy region of the signal splitter based receivers was evaluated. The numerical result told us that the power splitting based receiver outperforms its time switching based counterpart, when the linear RF-DC converter is conceived.

References

1. C. Cano, A. Pittolo, D. Malone, L. Lampe, A.M. Tonello, A.G. Dabak, State of the art in power line communications: from the applications to the medium. IEEE J. Sel. Areas Commun. **34**(7), 1935–1952 (2016)
2. J. Wu, H. Wu, C. Li, W. Li, X. He, C. Xia, Advanced four-pair architecture with input current balance function for power over ethernet (PoE) system. IEEE Trans. Power Electron. **28**(5), 2343–2355 (2013)
3. M. Gastpar, On capacity under receive and spatial spectrum-sharing constraints. IEEE Trans. Inf. Theory **53**(2), 471–487 (2007)
4. L.R. Varshney, Transporting information and energy simultaneously, in *2008 IEEE International Symposium on Information Theory* (2008), pp. 1612–1616
5. P. Grover, A. Sahai, Shannon meets tesla: wireless information and power transfer, in *2010 IEEE International Symposium on Information Theory* (2010), pp. 2363–2367
6. X. Zhou, R. Zhang, C.K. Ho, Wireless Information and power transfer: architecture design and rate-energy tradeoff. IEEE Trans. Commun. **61**(11), 4754–4767 (2013)
7. Z. Xiang, M. Tao, Robust beamforming for wireless information and power transmission. IEEE Wirel. Commun. Lett. **1**(4), 372–375 (2012)
8. B. Koo, D. Park, Interference alignment and wireless energy transfer via antenna selection. IEEE Commun. Lett. **18**(4), 548–551 (2014)
9. S. Zhao, Q. Li, Q. Zhang, J. Qin, Antenna selection for simultaneous wireless information and power transfer in MIMO systems. IEEE Commun. Lett. **18**(5), 789–792 (2014)
10. K. Liang, L. Zhao, Z. Ding, H.H. Chen, Double side signal splitting SWIPT for downlink CoMP transmissions with capacity limited backhaul. IEEE Commun. Lett. **20**(12), 2438–2441 (2016)
11. S. Li, W. Xu, Z. Liu, J. Lin, Independent power splitting for interference-corrupted SIMO SWIPT systems. IEEE Commun. Lett. **20**(3), 478–481 (2016)
12. G. Zhu, C. Zhong, H.A. Suraweera, G.K. Karagiannidis, Z. Zhang, T.A. Tsiftsis, Wireless information and power transfer in relay systems with multiple antennas and interference. IEEE Trans. Commun. **63**(4), 1400–1418 (2015)
13. L. Liu, R. Zhang, K.C. Chua, Wireless information and power transfer: a dynamic power splitting approach. IEEE Trans. Commun. **61**(9), 3990–4001 (2013)
14. R. Zhang, C.K. Ho, MIMO broadcasting for simultaneous wireless information and power transfer. IEEE Trans. Wirel. Commun. **12**(5), 1989–2001 (2013)
15. H. Liu, K.J. Kim, K.S. Kwak, H.V. Poor, Power splitting-based SWIPT with decode-and-forward full-duplex relaying. IEEE Trans. Wirel. Commun. **15**(11), 7561–7577 (2016)
16. Z. Chu, Z. Zhu, J. Hussein, Robust optimization for AN-aided transmission and power splitting for secure MISO SWIPT system. IEEE Commun. Lett. **20**(8), 1571–1574 (2016)
17. Y. Zhu, K.K. Wong, Y. Zhang, C. Masouros, Geometric power control for time-switching energy-harvesting two-user interference channel. IEEE Trans. Veh. Technol. **65**(12), 9759–9772 (2016)
18. Y. Dong, M.J. Hossain, J. Cheng, Joint power control and time switching for SWIPT systems with heterogeneous QoS requirements. IEEE Commun. Lett. **20**(2), 328–331 (2016)
19. S. Atapattu, H. Jiang, J. Evans, C. Tellambura, Time-switching energy harvesting in relay networks, in *2015 IEEE International Conference on Communications (ICC)* (2015), pp. 5416–5421
20. Y. Shen, X. Huang, K.S. Kwak, B. Yang, S. Wang, Subcarrier-pairing-based resource optimization for OFDM wireless powered relay transmissions with time switching scheme. IEEE Trans. Signal Process. **65**(5), 1130–1145 (2017)

Chapter 3
Throughput Maximization and Fairness Assurance in Data and Energy Integrated Communication Networks

Abstract In this chapter, we will study a typical data and energy integrated network (DEIN) having a single hybrid base station (H-BS), which is capable of simultaneously transmitting the data and energy to multiple user equipments (UEs) during the downlink transmissions. The UEs operating in this DEIN are capable of harvesting the energy from the H-BS's downlink transmissions by adopting the power splitting technique and they are also capable of exploiting the harvested energy for powering their uplink data transmissions. The H-BS's downlink transmissions and the UEs' uplink transmissions are time slotted in order to reduce the potential interference and transmission collision. Both of the uplink sum-throughput and the uplink fair-throughput of the DEIN will be maximised by determining the duration of each time-slot during the downlink and uplink transmissions and by determining the optimal power splitting factor for each UE. Both of these optimization problems are finally solved by the classic method of Lagrange multipliers in close-form. An interesting observation shows that supporting low-throughput data services during the DL transmissions does not degrade the wireless energy transfer and hence does not reduce the throughput of the UL transmissions.

Keywords Convex Optimisation · Data and Energy Integrated Communication Network (DEIN) · Fair-Throughput Maximisation · Lagrange Multiplier Medium-Access-Control (MAC) · Power Allocation · Power Splitting (PS) Resource Allocation · Simultaneous Wireless Information and Power Transfer Sum-Throughput Maximisation · Time-Division-Multiple-Access (TDMA) Wireless Powered Communication Network (WPCN) · Wireless Information and Power Transfer

In this chapter, we will study a typical data and energy integrated communication network (DEIN) having a single hybrid base station (H-BS), which is capable of simultaneously transmitting the data and energy to multiple user equipments (UEs) during the downlink transmissions by invoking the time-division-multiple-access (TDMA) protocol in the medium access control (MAC) layer. Specifically, each UE may receive a time slot during the downlink transmission in order to simultaneously receive the energy and the information from the H-BS, while each UE may also receive a time slot during the uplink transmission for the sake of uploading its own

J. Hu and K. Yang, *Data and Energy Integrated Communication Networks*, SpringerBriefs in Computer Science, https://doi.org/10.1007/978-981-13-0116-2_3

information to the H-BS. Then, both of the uplink sum-throughput and the uplink fair-throughput of the DEIN will be maximised by determining the duration of each time-slot for the downlink and uplink transmissions and by determining the optimal power splitting factor for each UE. Both of these optimization problems are finally solved by the classic method of Lagrange multipliers in close-form. An interesting observation shows that supporting low-throughput data services during the DL transmissions does not degrade the wireless energy transfer and hence does not reduce the throughput of the UL transmissions. We will discuss more details about the optimization issues in this chapter.

3.1 Introduction

Our cities now are in the process of transiting towards more smart, more automatic and more responsive societies, which requires the integration of the modern communication and information technology as well as the Internet of Things (IoT) [1]. The assets of smart cities contains smart transportation systems [2], smart grids [3], smart hospitals [4], smart factories [5] and etc. All these realisations require the universal connectivity of humans and machines. As foreseen by the industry, we will see more than 200 000 IoT devices deployed in a square kilometre.

Conventionally, the energy supplies of user equipments[1] (UEs) in wireless communication networks come from either batteries embedded or the power grid connected. However, these two energy sources have obvious limitations. The limited energy stored in the batteries restricts the lifetime of UEs, while the wire connected to the power grid restricts UEs' movement. Furthermore, massive IoT devices are deployed in walls or under roads or in other untouchable places. It is difficult to regularly replace their batteries, which limit their lifetime to a large extent. Embedding the function of energy harvesting (EH) into UEs and seeking energy from the renewable sources, such as sunlight [6] and wind [7], are capable of satisfying UEs' increasing energy demand [8]. However, energy arrivals from the renewable sources are stochastic processes, which hinders its efficient usage in supporting the communication functions of the UEs.

Transferring energy by radio frequency (RF) signals is more reliable and controllable than renewable energy sources. Zungeru et al. have demonstrated the availability of harvesting energy from the surrounding RF signals [9]. Varsheney has provided an information theoretical analysis for revealing the performance limit of simultaneous wireless information and power transfer (SWIPT) [10]. In order to process contaminated RF signals for the information reception as well as to convert RF signals into direct current (DC) for the energy harvesting, the spatial splitting [11], the power splitting [12] and the time switching [13] techniques have been invoked for the SWIPT. Many efforts then have been contributed to this prosperous subject [14–17]. However, most of them are merely based on the frequency-division-

[1]The advent of IoT redefines the concept of user equipments, which now includes both hand-held devices and machine-type devices.

multiple-access (FDMA) protocol, while assuming symmetric duration of the UEs' downlink and uplink transmissions. Their optimization formulation is inapplicable, when the time-division-multiple-access (TDMA) protocol is adopted in the MAC layer for supporting the information and energy transfer in the multi-user scenario, since their methodologies failed to optimize the durations of both the DL and the UL transmissions. Furthermore, wireless powered communication networks (WPCNs) relying on the TDMA protocol have been investigated in [18–20]. In WPCNs, the protocol of "harvest-then-transfer" is conceived [21]. As a result, UEs may harvest energy from the H-BS during their downlink transmissions, then the energy harvested by the UEs is exploited for supporting their uplink transmissions. However, in WPCNs, the downlink transmissions are dedicated to the wireless energy transfer. The simultaneous energy and data transfer has been largely ignored. Against this background, our novel contributions in this chapter are summarised as follows:

- A novel data and energy integrated communication network (DEIN) is systematically established. In this DEIN, the H-BS simultaneously transfers both of the information and the energy to the UEs by adopting the TDMA protocol in the MAC layer during its downlink transmissions. Then the UEs initiate their uplink data transmissions by exploiting the energy harvested during the downlink transmission stage.
- Relying on the tool of the convex optimization and the classic method of the Lagrange multipliers, the sum-throughput maximization problem for the uplink transmissions is solved by jointly optimizing the allocation of the time slots for both of the downlink and uplink transmissions and by optimising the power splitting factors of all the UEs.
- In order to further ensure the fairness among the UEs in the DEIN, the fair-throughput, which is defined as the minimum throughput among all the UEs' uplink transmissions, is also maximized by optimizing the allocation of the time slots for both of the downlink and uplink transmissions and by optimising the power splitting factors of all the UEs.

The rest of this chapter is organised as follows. Our DEIN model is introduced in Sect. 3.2, followed by the maximisation of the sum-throughput as well as the fair-throughput of the uplink transmissions in Sects. 3.3 and 3.4, respectively. Numerical results are provided in Sect. 3.5. Finally, we conclude this chapter in Sect. 3.6.

3.2 System Model

We consider a typical DEIN, as portrayed in Fig. 3.1, for the sake of remotely charging the UEs without violating their communication demands. The DEIN consists of a single H-BS as well as K UEs, which are denoted by the set $\{U_i | i = 1, \ldots, K\}$. The H-BS and the UEs are all equipped with a single antenna and they operate on the same spectral band, which indicates that the spatial and spectral resources are reused by

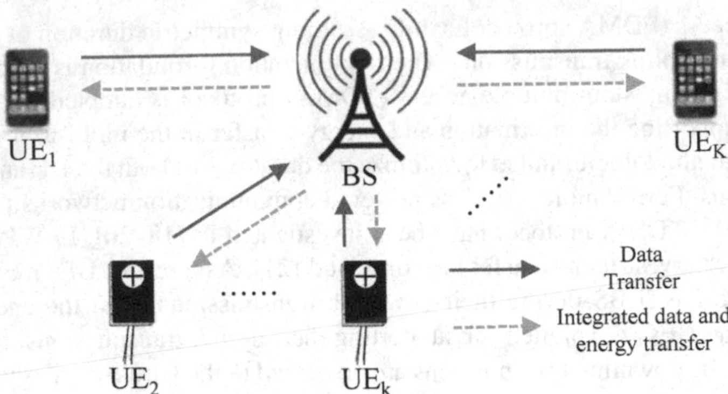

Fig. 3.1 The DL and UL transmissions of the DEIN

all the UEs. Moreover, the UEs in the DEIN are equipped with super capacitors [22]. Super capacitors may ideally store the energy that is gleaned from the RF signals without any energy loss. However, they suffer from low energy storage capacity and short discharging cycle. As a result, the UEs have to harvest energy from the downlink transmissions of the H-BS and store the energy in their super capacitors, while simultaneously recovering their requested data information from the same RF signals. The energy stored in the super capacitors is then depleted for powering the UEs' own uplink transmissions. We further assume that in this DEIN, all the channel state information is perfectly known by the H-BS.

3.2.1 Structure of the TDMA Aided Operating Cycle

In the DEIN studied, the UEs are fully powered by the energy gleaned from the RF signals of the H-BS's downlink transmission. As a result, their transmit power of the uplink transmissions is very low. The time-division-multiple-access (TDMA) protocol is then adopted in the MAC layer for reducing the hostile interference and transmission collision, when multiple UEs upload their own data to the H-BS. Furthermore, adopting the TDMA protocol during the downlink transmissions is also capable of orthogonally transmitting data to the multiple requesters with a substantial reduction of the interference. The UEs may also flexibly switch between the information decoding and energy harvesting operations in the power domain during the downlink transmissions of the H-BS.

The structure of a typical operating cycle having a duration of T is depicted in Fig. 3.2. An intact operating cycle consists of two phases, namely the control phase having a duration of T_{ctr} and the transmission phase having a duration of T_{tra}. During the control phase, the following tasks have to be completed by exchanging control signalling between the H-BS and the UEs:

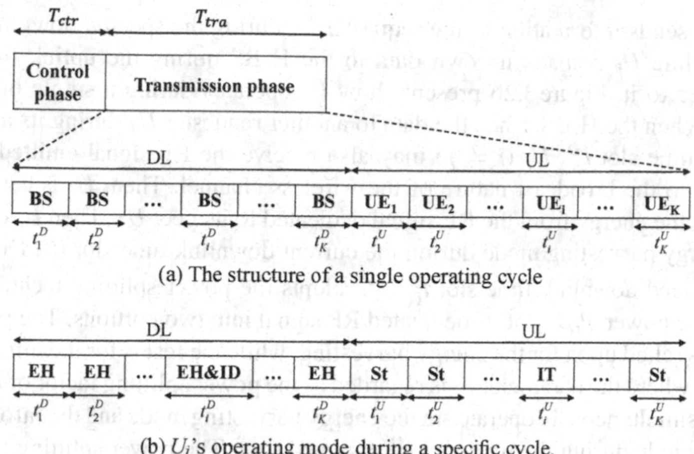

Fig. 3.2 Slotted downlink and uplink transmissions in the DEIN. EH: Energy Harvesting operation; ID: Information Decoding operation; St: Standby operation; IT: Information Transfer operation; DL: Downlink; UL: Uplink

- *Channel Estimation*: the channel state information can be acquired by the H-BS via the forward-link training together with the reverse-link feedback [23]. The channel states are assumed unchanged during a single operating cycle but they vary from one operating cycle to another. In this chapter, we simply assume that the H-BS is capable of acquiring the perfect channel state information.
- *Resource Allocation*: given the channel state information, the H-BS executes the time slot allocation for both the UEs' downlink and uplink transmissions and the H-BS also determines the signal splitting strategies at the UEs for their simultaneous data and energy reception. The H-BS then notifies the UEs about the time slot allocation scheme and the signal splitting strategies.
- *Synchronisation*: since all the UEs are distributed in the coverage of the H-BS, they may readily be synchronised together by invoking the time-stamp-based synchronization approach [24]. The H-BS may broadcast its locally recorded clock information to all the UEs during the control phase. Once the clock information is successfully received, the UEs may adjust their local clock in order to complete their synchronisation process.

The transmission phase of a single operating cycle is divided into a range of downlink time slots denoted by the set of $\mathbf{t}^D = \{t_i^D | i = 1, 2 \ldots, K\}$ and a range of uplink time slots denoted by the set of $\mathbf{t}^U = \{t_i^U | i = 1, 2 \ldots, K\}$. Hence, we have the following inequality, which is expressed as

$$\sum_{i=1}^{K} (t_i^D + t_i^U) \leq T_{tra}. \tag{3.1}$$

The H-BS sends information to the requester U_i during the specific downlink time slot t_i^D, while U_i uploads its own data to the H-BS during the uplink time slot t_i^U assigned to it. Figure 3.2b presents how U_i operates during a single operating cycle T. When the H-BS sends the data to another requester U_j during its assigned downlink time slot t_j^D, U_i $(i \neq j)$ may also receive the RF signal emitted by the H-BS due to the broadcast nature of the wireless channel. Then, U_i is capable of harvesting the energy from the RF signal dedicated to its peer U_j. Then U_i operates in the energy harvesting mode during the current downlink time slot t_j^D. During its own dedicated downlink time slot t_i^D, U_i adopts the power splitting technique for splitting the power $P_{recv,i}$ of its dedicated RF signal into two portions. The power of $\rho_i P_{recv,i}$ is relied upon for the energy harvesting, while the rest is for the information decoding, where the parameter ρ_i is regarded as the power splitting factor of U_i. As a result, U_i simultaneously operates in the energy harvesting mode and the information decoding mode during the time slot t_i^D assigned to it. The power splitting factor ρ_i can be adjusted by U_i in order to satisfy different energy and data requirements. For the uplink transmission, since only a single UE is allowed to transfer its data during a specific time slot, U_i solely operates in the information transfer mode during its assigned uplink time slot t_i^U. By contrast, U_i operates in the standby mode during other uplink time slots $\{t_j^U | j \neq i\}$ in order to avoid any transmission collision, when the corresponding UE U_j operates in the information transfer mode.

3.2.2 Channel Model

The downlink channel from the H-BS to U_i and the corresponding reversed uplink channel are denoted by the complex random variables $\tilde{h}_i$ and $\tilde{g}_i$, respectively, while their power coefficients are denoted by $h_i = |\tilde{h}_i|^2$ and $g_i = |\tilde{g}_i|^2$. For simplicity, we assume a symmetric channel between the H-BS and U_i, which indicates that $h_i = g_i$. The set of channel power coefficients is denoted as $\mathbf{h} = \{h_i | i = 1, \ldots, K\}$. Furthermore, the uncorrelated block fading channel models are conceived, which indicates that the power coefficient of the channel remain unchanged during a single operating cycle T. The channel noise power is denoted by $\sigma_{c,i}^2$, while the noise power of the information decoder is denoted by $\sigma_{ID,i}^2$. Compared to $\sigma_{ID,i}^2$, the channel noise power $\sigma_{c,i}^2$ is negligibly small and hence it has little influence on both of the practical information decoding and the energy harvesting [12]. As a result, the channel noise power $\sigma_{c,i}^2$ can be reasonably ignored in any of the formulations below. Furthermore, the noise power of the information decoder is assumed to be identical for every UE as well as the H-BS. For simplicity, we let $\sigma^2 = \sigma_{ID,i}^2$ denote the noise power of the information decoder in the following problem formulation.

3.2.3 Throughput of the Downlink Transmission

During the downlink time slot t_i^D, the power of the RF signal received by U_i is denoted by $P_{recv,i} = P_{BS}h_i$, where P_{BS} is the transmit power of the H-BS.

Since only a fraction of the received signal power is exploited by U_i for the information decoding, the achievable downlink throughput R_i^D of U_i can be expressed as the following formula by exploiting the classic Shannon's channel capacity equation:

$$R_i^D(t_i^D, \rho_i) = t_i^D \log_2[1 + \frac{(1-\rho_i)P_{recv,i}}{\sigma^2}], \text{ [bit/Hz]}, \tag{3.2}$$

The downlink throughput of (3.2) can also be regarded as the bandwidth efficiency of the DL data transfer. Therefore, the bandwidth term of the classic Shannon's channel capacity equation is not included in (3.2).

3.2.4 Throughput of the Uplink Transmission

The total energy harvested by U_i is the sum of the energy harvested during the downlink time slot set $\{t_j^D | j \neq i\}$ of its peers, when U_i operates in the energy harvesting mode, and the energy harvested during its dedicated downlink time slot t_i^D, when U_i operates in both of the energy harvesting and information decoding modes simultaneously. The total energy harvested by U_i can then be further expressed as

$$E_{recv,i}(\mathbf{t}^D, \rho_i) = \beta_i P_{recv,i}(\sum_{j \neq i} t_j^D + t_i^D \rho_i), \tag{3.3}$$

where β_i represents the efficiency of converting the alternative current (AC) energy carried by the RF signal to the DC that can be exploited for driving any electronic load. Here, for simplicity, the energy conversion efficiency β_i is assumed to be a unity.

Since the energy harvested by U_i during the H-BS's downlink transmission is fully exploited for powering its own uplink transmission, with the aid of (3.3), the achievable uplink throughput R_i^U of U_i can then be formulated as

$$\begin{aligned} &R_i^U(\mathbf{t}^D, t_i^U, \rho_i) \\ &= t_i^U \log_2 \left[1 + \frac{h_i P_{recv,i}(\sum_{j \neq i} t_j^D + t_i^D \rho_i)}{t_i^U \sigma^2} \right], \text{ [bit/Hz]}, \end{aligned} \tag{3.4}$$

which can also be regarded as the bandwidth efficiency of the uplink data transfer.

In our model, the diverse minimum throughput requirements of the downlink transmissions requested by the UEs can be represented by the set $\mathbf{D} = \{D_1, \ldots, D_K\}$. Our ultimate objective is to maximize the throughput of the uplink transmissions subject to the constraint that every UE's achievable downlink throughput should satisfy its minimum requirement by jointly optimizing the durations of the time slots in the downlink set $\mathbf{t}^D$ and those of the time slots in the uplink set $\mathbf{t}^U$ as well as the signal splitting strategies adopted by the UEs during their dedicated downlink time slots. The signal splitting strategies are represented by the power splitting factors in the set $\rho = \{\rho_i | 1 \leq \rho_i \leq K\}$. Furthermore, our model focuses on both of the sum-throughput maximisation for achieving the upper-bound of the UEs' uplink transmissions and the fair-throughput maximisation for ensuring the UEs' fairness during their uplink transmissions.

3.3 Sum-Throughput Maximisation

In this section, the sum-throughput maximization problem is formulated, and then it is transformed into a convex problem, which can be solved by the classic method of the Lagrange multipliers. With the aid of Eqs. (3.1)–(3.4), the sum-throughput maximization problem can be formulated as

$$(\text{P1}): \max_{\mathbf{t}^D, \mathbf{t}^U, \rho} \quad \sum_{i=1}^{K} R_i^U(\mathbf{t}^D, t_i^U, \rho_i) \tag{3.5}$$

$$\text{s.t.} R_i^D(t_i^D, \rho_i) \geq D_i, \tag{3.5a}$$

$$\sum_{i=1}^{K} (t_i^D + t_i^U) \leq T_{tra}, \tag{3.5b}$$

$$0 \leq \rho_i \leq 1, \tag{3.5c}$$

where $i = 1, \ldots, K$ denotes the indices of the UEs. Since $R_i^U(\mathbf{t}^D, t_i^U, \rho_i)$ of (3.4) and $R_i^D(t_i^D, \rho_i)$ of (3.2) are neither convex nor concave functions according to the definition of convexity, (P1) is thus a non-convex problem with respect to the variables $\mathbf{t}^D$, $\mathbf{t}^U$ and ρ. As a result, (P1) has to be equivalently transformed into a convex problem by introducing a new set of variables $\mu = \{\mu_i | i = 1, \cdot, K\}$ for substituting the original set of variables $\rho = \{\rho_i | i = 1, \ldots, K\}$. The i-th entry μ_i is then expressed as

$$\mu_i = t_i^D \rho_i, i = 1, \ldots, K. \tag{3.6}$$

Accordingly, the expression of the achievable downlink throughput R_i^D of U_i during its dedicated downlink time slot t_i^D can be reformulated as

$$R_i^D(t_i^D, \mu_i) = t_i^D \log_2(1 + \gamma_i - \gamma_i \frac{\mu_i}{t_i^D}), \tag{3.7}$$

where $\gamma_i = \frac{P_{recv,i}}{\sigma_2}$ for all $i = 1, \ldots, K$ represents the signal-to-noise-ratio (SNR) of U_i during the time slot t_i^D. The set of the UEs' received SNRs during the H-BS's downlink transmissions is denoted as $\gamma = \{\gamma_i | i = 1, \ldots, K\}$. The expression of the achievable uplink throughput R_i^U of U_i during its assigned uplink time slot t_i^U can be further derived as

$$R_i^U(\mathbf{t}^D, t_i^U, \mu_i) = t_i^U \log_2(1 + \frac{h_i \gamma_i (\sum_{j \neq i} t_j^D + \mu_i)}{t_i^U}), \tag{3.8}$$

while the power splitting factor of U_i during its assigned downlink time slot t_i^D can be expressed as

$$\rho_i = \frac{\mu_i}{t_i^D}. \tag{3.9}$$

Therefore, the original optimisation problem (P1) can be reformulated as

$$(\text{P2}): \max_{\mathbf{t}^D, \mathbf{t}^U, \mu} \quad \sum_{i=1}^{K} R_i^U(\mathbf{t}^D, t_i^U, \mu_i), \tag{3.10}$$

$$\text{s.t.} \quad R_i^D(t_i^D, \mu_i) \geq D_i, \tag{3.10a}$$

$$\sum_{i=1}^{K} (t_i^D + t_i^U) \leq T_{tra}, \tag{3.10b}$$

$$0 \leq \mu_i \leq t_i^D, \tag{3.10c}$$

where $i = 1, \ldots, K$. Since $f(\mathbf{t}^D, \mu_i) = \log_2[1 + h_i \gamma_i (\sum_{j \neq i} t_j^D + \mu_i)]$ is a concave function, its log-affine $R_i^U(\mathbf{t}^D, t_i^U, \mu_i)$ is concave as well. Therefore, the objective function (3.10) of the alternative optimisation problem (P2), which is the sum of a range of concave functions, can be readily proved to be concave with respect to the variables $\mathbf{t}^D$, $\mathbf{t}^U$ and μ. Furthermore, $R_i^D(t_i^D, \mu_i)$ in (3.10a) is also a concave function of these deciding variables since its Hessian matrix is positive semi-definite, while the constrains (3.10b) and (3.10c) are both affine. As a result, (P2) is a convex optimization problem.

Observe from the optimisation problem (P2) that the downlink transmission requirement D_i of U_i should be higher than zero and smaller than its maximum achievable downlink throughput R_i^D, when U_i exploits all its received RF signal for the information decoding by completely sacrificing its energy harvesting function during its dedicated downlink time slot t_i^D. The Lagrange function of (P2) can be then formulated as

$$\mathcal{L}(\mathbf{t}^D, \mathbf{t}^U, \mu, \lambda, \xi) = \sum_{i=1}^{K} R_i^U(\mathbf{t}^D, t_i^U, \mu_i) + \lambda[T_{tra} - \sum_{i=1}^{K}(t_i^D + t_i^U)] + \sum_{i=1}^{K} \xi_i[R_i^D(t_i^D, \mu_i) - D_i], \quad (3.11)$$

where λ and $\xi = \{\xi_i | i = 1, \ldots, K\}$ are the corresponding Lagrangian multipliers. Moreover, the dual function of (P2) can be then expressed as

$$\mathcal{G}(\lambda, \xi) = \sup \mathcal{L}(\mathbf{t}^D, \mathbf{t}^U, \mu, \lambda, \xi). \quad (3.12)$$

Since (P2) is a convex optimization problem, its optimal solutions, which is denoted as $\{\mathbf{t}^{D*}, \mathbf{t}^{U*}, \mu^*, \lambda^*, \xi^*\}$, have to satisfy the following Karush-Kuhn-Tucker (KKT) conditions:

$$\ln(1 + y_i) - \frac{y_i}{1 + y_i} = \lambda \ln 2, \quad (3.13)$$

$$\sum_{j \neq i} \frac{\gamma_j h_j}{1 + y_j} + \xi_i[\ln(1 + \gamma_i - z_i) + \frac{z_i}{1 + r_i - z_i}] = \lambda \ln 2, \quad (3.14)$$

$$\frac{\gamma_i h_i}{1 + y_i} = \xi_i \frac{\gamma_i}{1 + \gamma_i - z_i}, \quad (3.15)$$

$$\lambda[T_{tra} - \sum_{i=1}^{K}(t_i^U + t_i^D)] = 0, \quad (3.16)$$

$$\xi_i[t_i^D \log_2(1 + \gamma_i - z_i) - D_i] = 0, \quad (3.17)$$

where we introduce a couple of new variables sets, denoted by $\mathbf{y} = \{y_i | i = 1, \ldots, K\}$ and $\mathbf{z} = \{z_i | i = 1, \ldots, K\}$. Their i-th entries can be expressed as

$$y_i = h_i \gamma_i \frac{\sum_{j \neq i} t_j^D + \mu_i}{t_i^U}, \quad (3.18)$$

$$z_i = \gamma_i \frac{\mu_i}{t_i^D}, \quad (3.19)$$

respectively, for $i = 1, \ldots, K$. According to (3.14) and (3.15), we can find that $\lambda \neq 0$ and $\xi_i \neq 0$.

Given a specific value of the Lagrange multiplier λ and according to (3.13)–(3.19), the resultant optimal value t_i^{D*} of the duration of downlink time slot assigned to U_i

can be derived as

$$t_i^{D*} = \frac{D_i}{\log_2(1 + \gamma_i - z_i^*)}. \tag{3.20}$$

Furthermore, the optimal value t_i^{U*} of the duration of the uplink time slot assigned to U_i can be obtained as

$$t_i^{U*} = h_i \gamma_i \frac{\sum_{j \neq i} t_j^{D*} + \mu_i^*}{y_i^*}, . \tag{3.21}$$

The optimal value of the intermediate variable μ_i^* can be calculated as

$$\mu_i^* = \frac{z_i^* t_i^{D*}}{\gamma_i}. \tag{3.22}$$

In Eqs. (3.20), (3.21) and (3.22), y_i^* and z_i^* are the solutions to the following equations:

$$\ln(1 + y_i) - \frac{y_i}{1 + y_i} = \lambda \ln 2, \tag{3.23}$$

$$h_i(1 + \gamma_i - z_i)\ln(1 + \gamma_i - z_i) + h_i z_i = (1 + y_i)\lambda \ln 2 - \sum_{j \neq i} \gamma_j h_j. \tag{3.24}$$

The expression on the left side of (3.23) increases monotonically with respect to the variable y_i, while the expression on the left side of equation (3.24) decreases monotonically with respect to the variable z_i. As a result, y_i^* can be calculated first by invoking the classic bisection method. Substituting y_i^* into (3.24), z_i^* can also be calculated by invoking the classic bisection method.

Given the specific value of λ, we have obtained the optimal values of $\mathbf{t}^{U*}$, $\mathbf{t}^{D*}$, μ^* and ξ^*, which satisfy the equalities of (3.13)–(3.15) and (3.17). Then, the sub-gradient descent is invoked for iteratively obtaining the optimal Lagrange multiplier λ^*. The sub-gradient of $\mathcal{G}(\lambda, \xi)$ with respect to the Lagrange multiplier λ, which is denoted by $p(\lambda)$, can be further expressed as

$$p(\lambda) = T_{tra} - \sum_{i=1}^{K} (t_i^D + t_i^U). \tag{3.25}$$

With the aid of (3.25), we can iteratively obtain the optimal Lagrange multiplier λ^* by obeying the following steps. We update λ by the formula $\lambda^{(n)} = \lambda^{(n-1)} - p(\lambda^{(n-1)})\Delta_\lambda$ in each iteration, where n denotes the n-th iteration and Δ_λ represents the step length of each iteration. Substituting $\lambda^{(n)}$ into (3.20)–(3.24), we may obtain the corresponding values of t_i^D and t_i^U and hence derive the specific value of $p(\lambda^{(n)})$. The iteration continues until we find the optimal λ^*, which makes $|p(\lambda^*)|$ smaller than the specific error tolerance δ. Finally, the power splitting factor set ρ^* can be calculated by invoking (3.9). The procedure of iteratively solving the optimisation problem (P2) is detailed in the pseudo code of Algorithm 1.

Algorithm 1 Iterative algorithm for solving (P2)

Require: Duration of the transmission phase T_{tra}; DL throughput requirement **D**; channel power gains **h**; SNR in UE $\boldsymbol{\gamma}$; error tolerance δ

Ensure: optimal allocated UL time slots $\mathbf{t}^{D*}$; optimal allocated DL time slots $\mathbf{t}^{U*}$; optimal PS factors $\boldsymbol{\rho}^*$;

```
Transform (P1) to (P2) by substituting μ for ρ;
Initialize λ > 0 and iteration step length Δ_λ > 0 and p(λ) > δ;
while |p(λ)| > δ do
    Calculate y* and z* by equations (3.23), (3.24);
    Calculate t^{D*}, μ* and t^{U*} by equations (3.20)-(3.21);
    Update p(λ) by equation (3.25);
    Update λ by λ = λ − p(λ)Δ_λ;
end while
Calculate ρ* by equation (3.9);
return t^{D*}, t^{U*}, ρ*
```

3.4 Fair-Throughput Maximisation

In order to achieve a better sum-throughput, more resources are inclined to be allocated to the UEs having better channel qualities between the H-BS. Since the channel qualities are largely determined by the large-scale channel attenuation, such as the path-loss, the UEs close to the H-BS may gain more resources for harvesting energy from the H-BS's downlink transmission and for uploading their own data to the H-BS during their uplink transmissions. As a result, the UEs relatively far away from the H-BS may not be allocated sufficient resources for their own operations. This is regarded as the classic near-far effect, which yields the fairness issue among the UEs in the DEIN.

In order to overcome the classic near-far effect during the resource allocation, ensuring the fairness among the UEs' uplink transmissions becomes our prim objective, which yields the maximization of the so-called fair-throughput. Since fair-throughput represents the minimum throughput among all the UEs during their uplink transmissions, we impose a constraint on the throughput of the UEs' uplink transmissions, which is expressed as $R_i^U(\mathbf{t}^D, t_i^U, \rho_i) \geq R$, for $i = 1, \ldots, K$, where R represents the so-called fair-throughput. According to the system model of Sect. 3.2, the fair-throughput maximization problem (P3) can be formulated as

$$\text{(P3)}: \max_{\mathbf{t}^D, \mathbf{t}^U, \boldsymbol{\mu}} \quad R \tag{3.26}$$

$$\text{s.t.} \quad R_i^D(t_i^D, \mu_i) \geq D_i, \tag{3.26a}$$

$$R_i^U(\mathbf{t}^D, t_i^U, \mu_i) \geq R, \tag{3.26b}$$

$$0 \leq \mu_i \leq t_i^D, \tag{3.26c}$$

$$\sum_{i=1}^{K} (t_i^D + t_i^U) \leq T_{tra}, \tag{3.26d}$$

where $\mu_i = t_i^D \rho_i$ for all $i = 1, \ldots, K$ is adopted for ensuring the concavity of both the achievable downlink throughput $R_i^D(t_i^D, \mu_i)$ of U_i during its downlink time slot t_i^D and the achievable UL throughput $R_i^U(\mathbf{t}^\mathbf{D}, t_i^U, \mu_i)$ during its uplink time slot t_i^U, which have been proved in Sect. 3.3. As a result, the fair-throughput maximisation problem (P3) can be readily proved to be a convex optimization problem. Note that the achievable downlink throughput $R_i^D(t_i^D, \mu_i)$ is an increasing function with respect to t_i^D, while the achievable uplink throughput $R_i^U(\mathbf{t}^D, t_i^U, \mu_i)$ is also an increasing function with respect to t_i^D and t_i^U. Therefore, the fair-throughput R increases when $t = \sum_{i=1}^{K} t_i^D + t_i^U$ increases. As a result, we may iteratively solve the following convex optimisation problem (P4) in order to maximize the fair-throughput R:

$$(\text{P4}): \min_{\mathbf{t}^D, \mathbf{t}^U, \mu} \quad \sum_{i=1}^{K} (t_i^D + t_i^U) \tag{3.27}$$

$$\text{s.t.} \quad R_i^D(t_i^D, \mu_i) \geq D_i, \tag{3.27a}$$

$$R_i^U(\mathbf{t}^D, t_i^U, \mu_i) \geq R, \tag{3.27b}$$

$$0 \leq \mu_i \leq t_i^D, \tag{3.27c}$$

where $i = 1, \ldots, K$. The Lagrange function of (P4) is further expressed as

$$\mathcal{L}(\mathbf{t}^D, \mathbf{t}^U, \mu, \lambda, \xi) = \sum_{i=1}^{K} (t_i^D + t_i^U) + \sum_{i=1}^{K} \xi_i [D_i - R_i^D(t_i^D, \mu_i)] + \sum_{i=1}^{K} \lambda_i [R - R_i^U(\mathbf{t}^D, t_i^U, \mu_i)], \tag{3.28}$$

where $\lambda = \{\lambda_i | i = 1, \ldots, K\}$ and $\xi = \{\xi_i | i = 1, \ldots, K\}$ are the corresponding Lagrangian multipliers. The dual function of (P4) then can be expressed as

$$\mathcal{G}(\lambda, \xi) = \inf \mathcal{L}(\mathbf{t}^D, \mathbf{t}^U, \mu, \lambda, \xi). \tag{3.29}$$

Similar to the method invoked for solving the sum-throughput maximisation problem (P2), the KKT conditions are also exploited for solving the fair-throughput maximisation problem (P4). Hence, given a range of specific values for the multiplier set $\lambda = \{\lambda_i | i = 1, \ldots, K\}$, the optimal value of the duration of the DL time slot t_i^{D*}, that of the duration of the UL time slot t_i^{U*} and that of the intermediate variable μ_i^* can still be expressed by (3.20)–(3.22). Furthermore, y_i^* and z_i^* can be obtained by solving the following equations:

$$\ln(1+y_i) - \frac{y_i}{1+y_i} = \frac{\ln 2}{\lambda_i}, \tag{3.30}$$

$$(1+\gamma_i - z_i)\ln(1+\gamma_i - z_i) + z_i = \frac{1+y_i}{\lambda_i h_i}\ln 2 - \frac{\sum_{j \neq i} \lambda_j \gamma_j h_j}{\lambda_i h_i}. \tag{3.31}$$

Relying on the monotonous properties of the expressions on the left side of the equalities of (3.30) and (3.31), we can readily obtain the solutions of y_i^* and z_i^* by invoking the classic bisection method. The sub-gradient of $\mathcal{G}(\lambda, \xi)$ with respect to λ, which is denoted by $p(\lambda) = \{p(\lambda_i) | i = 1, \ldots, K\}$, can be further expressed as

$$p(\lambda_i) = t_i^{U*}\log_2(1+y_i^*) - R, \tag{3.32}$$

for all $i = 1, \ldots, K$. We iteratively update the Lagrange multiplier set λ by $\lambda^{(n)} = \lambda^{(n-1)} - p(\lambda)\Delta_\lambda$ in each iteration, where n denotes the n-th iteration and Δ_λ represents the step length of the iteration. The iteration for obtaining the optimal Lagrange multiplier set λ^* terminates until the sub-gradient of the dual function $\mathcal{G}(\lambda, \xi)$ satisfies the condition of $|p(\lambda^*)| \leq \delta_\lambda$, where δ_λ represents the absolute error tolerance of the Lagrange multiplier set λ.

We reduce the fair-throughput R after obtaining the optimal result t^* by solving the alternative optimisation problem (P4), if the optimal result t^* is higher than the duration T_{tra} of the transmission phase, say $t^* > T_{tra}$. By contrast, if the optimal result t^* is lower than the duration T_{tra} of the transmission phase, say $t^* < T_{tra}$, we have to increase the fair-throughput R. This iteration process terminates until we have $|T_{tra} - t^*| < \delta_R$, which yields the maximum fair-throughput R^*. Here, δ_R represents the error tolerance. The iterative algorithm of solving the optimisation problem (P4) is detailed in Algorithm 2.

3.5 Numerical Results

In this section, the numerical results of the maximum sum-throughput obtained by solving the optimisation problem (P1) and those of the maximum fair-throughput obtained by solving the optimisation problem (P3) are compared with each other in a typical DEIN consisting of a H-BS and several UEs. Without loss of generality,

Algorithm 2 Iterative algorithm for solving (P4))

Require: duration of the transmission phase T_{tra}; DL throughput requirement $\mathbf{D}$; channel power gains $\mathbf{h}$; SNR in UE γ; error tolerance δ_λ and δ_R

Ensure: optimal allocated UL time slots $\mathbf{t}^{D*}$; optimal allocated DL time slots $\mathbf{t}^{U*}$; optimal PS factors $\boldsymbol{\rho}^*$; optimal fair-throughput R^*

```
Initialize $R_{min} = 0$ and $R_{max}$ (large enough) and $t^* = 0$;
while $|T_{tra} - t^*| > \delta_R$ do
  Let $R = 0.5(R_{max} + R_{min})$;
  Initialize $\lambda_i > 0$ and $\Delta_\lambda > 0$ and $p(\boldsymbol{\lambda})$ (let $|p(\boldsymbol{\lambda})| > \delta_\lambda$);
  while $|p(\boldsymbol{\lambda})| > \delta_\lambda$ do
    Calculate $\mathbf{y}^*$ and $\mathbf{z}^*$ by equations (3.30) and (3.31);
    Calculate $\mathbf{t}^{D*}$, $\mathbf{t}^{U*}$ $\boldsymbol{\mu}^*$ by equations (3.20)-(3.21);
    Calculate $p(\boldsymbol{\lambda}) = \{p(\lambda_i)|i = 1, \cdots, K\}$ by equation (3.32);
    Update $\boldsymbol{\lambda}$ by $\boldsymbol{\lambda} = \boldsymbol{\lambda} - p(\boldsymbol{\lambda})\Delta_\lambda$;
  end while
  Calculate $t^* = \sum_{i=1}^{K} t_i^{D*} + t_i^{U*}$;
  if $|T_{tra} - t^*| > \delta_R$ then
    if $t^* > T_{tra}$ then
      Let $R_{max} = R$;
    else
      Let $R_{min} = R$;
    end if
  end if
end while
Calculate $\boldsymbol{\rho}^*$ by equation (3.9);
return $\mathbf{t}^{D*}$, $\mathbf{t}^{U*}$, $\boldsymbol{\rho}^*$,$R^*$
```

the Additive-White-Gaussian-Noise (AWGN) channel as well as the path loss are conceived. Therefore, the DL and UL channel power gains are modelled by $h_i = g_i = 10^{-3}Y_i^{-\alpha}$, for all $i = 1, \ldots, K$, where Y_i represents the distance between the H-BS and U_i. The exponent is set to be $\alpha = 2$ for representing the short-range free-space path loss model. A 30 dB signal power attenuation in average is assumed at a reference distance of 1 m for this channel model. The noise power of the information decoder is set to be −50 dBm, while the channel noise is ignored.

We first compare the UE's individual uplink throughput obtained by solving the sum-throughput maximization problem (P1) to that obtained by solving the fair-throughput maximization problem (P3). The transmit power P_{BS} is set to be 30 dBm. We have $K = 5$ UEs in total in the DEIN. The distances from the UEs to the H-BS are $\{Y_1 = 4, Y_2 = 5, Y_3 = 5.5, Y_4 = 9, Y_5 = 10\}$ m, while the minimum requirements of the UEs' downlink throughput are $\{D_1 = 0.5, D_2 = 0.4, D_3 = 0.8, D_4 = 0.3, D_5 = 0.2\}$ bit/Hz. The duration of the transmission phase is $T_{tra} = 1$ s.

As illustrated in Fig. 3.3, the UEs within the proximity of the H-BS, such as U_1 and U_2, are capable of transferring more data during their uplink transmissions than the UEs far away from the H-BS, such as U_4 and U_5, if we aim for maximizing the sum-throughput of the UEs' uplink transmissions. In order to achieve this objective, more time resources are assigned to the UEs having better channel qualities, which results in the substantial unfairness among the UEs. As a result, in order to attain a fair resource allocation scheme, the maximization of the fair-throughput is studied in order to ensure the fairness among the UEs by suffering somewhat degradation of the sum-throughput. We can observe from Fig. 3.3 that in order to maximize the fair-throughput, the actual UL throughput of different UEs are soundly fair by allocating more time resources to the UEs having worse channel qualities for the sake of overcoming the adverse near-far effect.

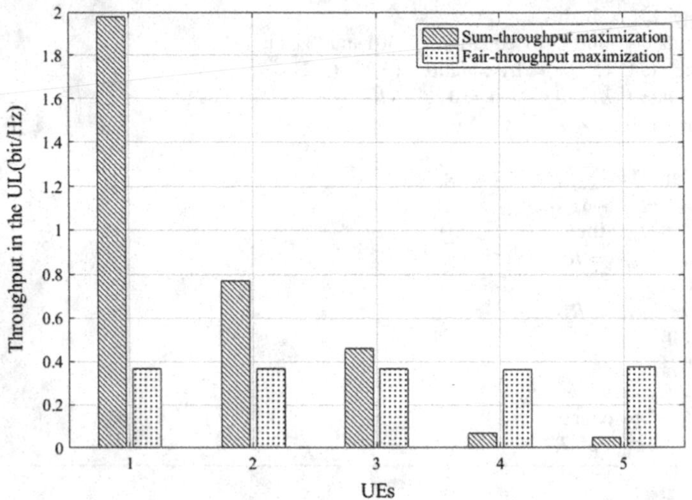

Fig. 3.3 The individual throughput of the UEs' uplink transmissions

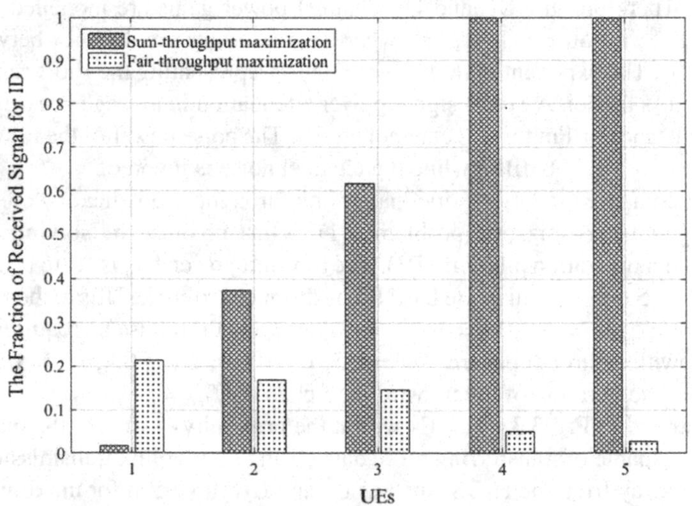

Fig. 3.4 The power splitting strategies of the UEs during the H-BS's downlink transmissions

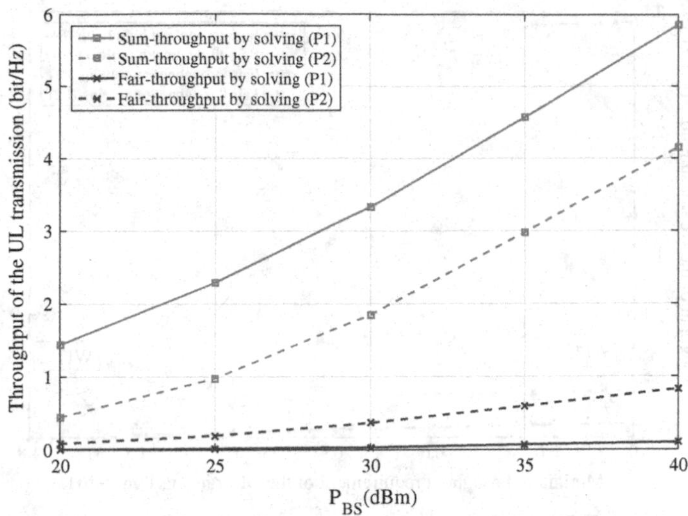

Fig. 3.5 Throughput of the UL transmission versus the transmit power of the H-BS

Furthermore, we plot the power splitting strategies for these five UEs during the H-BS's downlink transmissions in Fig. 3.4. We first focus on the sum-throughput maximisation. Since U_1 and U_2 are very close to the H-BS, they only exploit a very small fraction of their received signal for the information decoding so as to satisfy their downlink throughput requirement. The rest of their received signal is all converted to the DC energy, which is exploited for powering their uplink transmissions. As a result, U_1 and U_2 are capable of achieving higher uplink transmission throughput. By contrast, since U_4 and U_5 are far away from the H-BS, they have to exploit all their received signals for the information decoding. As a result, they do not harvest sufficient energy for powering their uplink transmissions. Hence, they suffer from very low uplink transmission throughput. Note that when the fair-throughput maximisation is invoked in our resource allocation and power splitting strategy selection schemes, all the UEs choose moderate power splitting strategies in order to achieve the fairness of their uplink transmissions.

We further plot both of the sum-throughput and the fair-throughput against the transmit power P_{BS} of the H-BS in Fig. 3.5, where we adopt the same parameter setting as those for obtaining the numerical results of Fig. 3.3. Observe from Fig. 3.5 that when the transmit power P_{BS} of the H-BS increases, both of the sum-throughput obtained by solving the optimisation problem (P1) and the fair-throughput obtained by solving the optimisation problem (P3) increase. Furthermore, the sum-throughput is more sensitive to the increase of P_{BS} than the fair-throughput. Without considering the fairness among the UEs, the uplink throughputs of the UEs near the H-BS may be significantly increased by increasing the transmit power of P_{BS}. However, the uplink throughputs of the UEs far away from the H-BS may be improved little due to the signal propagation of long distances. Hence, the substantial increase of the

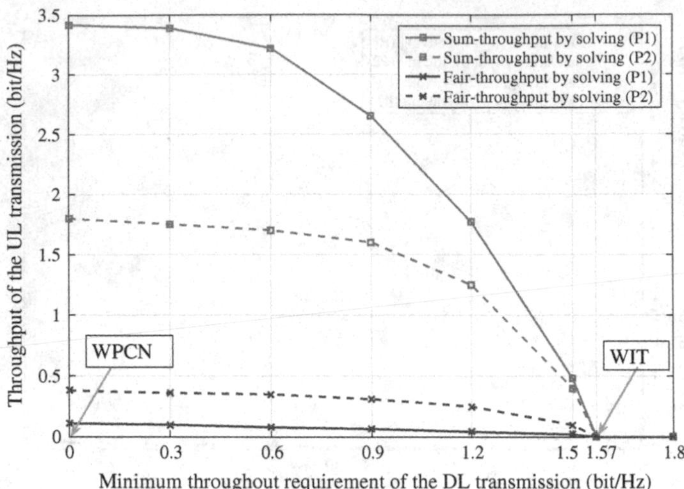

Fig. 3.6 Throughput of the UL transmission versus the minimum throughput requirement in the DL transmission

sum-throughput is mainly contributed by the UEs near the H-BS. Since the fair-throughput mainly depends on the UEs having worse channel qualities, it may not be improved a lot by increasing the transmit power P_{BS} of the H-BS.

We plot both of the sum-throughput and the fair-throughput of the UEs' uplink transmissions against the minimum throughput requirement of the downlink transmissions in Fig. 3.6. For simplicity, we set the minimum downlink throughput requirement identical for all the UEs. Note that if the minimum downlink throughput requirement falls to zero, our DEIN becomes a typical WPCN. In the WPCN, the downlink transmission of the H-BS does not carry any requested data. The downlink signal is only for transferring the energy to the UEs. Observe from Fig. 3.6 that as we increase the minimum downlink throughput requirement, both of the sum-throughput and fair-throughput of the uplink transmissions are reduced. We also have an interesting observation that when the minimum downlink throughput requirement is lower than 0.3 bit/Hz, it has little influence on both of the sum-throughput and the fair-throughput of the uplink transmissions. The observation indicates that our DEIN can efficiently support the low-rate downlink transmission, such as the signalling exchange, while fulfilling the wireless charging tasks, without any significant loss of the UEs' UL transmissions. As the minimum throughput requirement of the downlink transmission continually increases, both of the sum-throughput and the fair-throughput gradually become zero. This is because a large portion of the received RF signal of the downlink transmission is fully exploited for the information decoder in order to satisfy the harsh downlink throughput requirement and hence the UEs cannot harvest sufficient energy for supporting their own UL transmissions. As shown in Fig. 3.6, the uplink throughput reduces to zero, when the minimum downlink throughput requirement increases to 1.57 (bit/Hz). At this moment, the UEs completely sacrifice the function of the energy harvesting in order to achieve the minimum downlink throughput requirements, which makes our DEIN model a conventional wireless information transfer (WIT) system. If the downlink throughput requirement is higher than 1.57 (bit/Hz), this requirement is beyond the transmission capability of the DEIN in the current parameter settings.

3.6 Summary

In this chapter, we have studied a novel DEIN model, where the H-BS simultaneously transmit the data and energy during the downlink transmissions and the UEs harvest the energy from the downlink signals for powering their own uplink transmissions. In order to reduce potential collision and interference, a TDMA protocol is adopted in the MAC layer for both the downlink and uplink transmissions. At a UE's end, the received RF signal is split in the power domain. One portion of the signal is for the information decoding, while the other is for the energy harvesting. Relying on the classic convex optimization theory, both of the sum-throughput and the fair-throughput are maximised by optimizing both of the time slots allocation and the power spliting factors. Iterative algorithms are proposed for numerically solving the throughput maximization problems. Furthermore, our numerical results demonstrate the advantage of our DEIN over the WPCN and the WIT systems.

References

1. W. Ejaz, M. Naeem, A. Shahid, A. Anpalagan, M. Jo, Efficient energy management for the internet of things in smart cities. IEEE Commun. Mag. **55**(1), 84–91 (2017)
2. F. Zhu, Z. Li, S. Chen, G. Xiong, Parallel transportation management and control system and its applications in building smart cities. IEEE Trans. Intell. Transp. Syst. **17**(6), 1576–1585 (2016)
3. A. Boustani, A. Maiti, S.Y. Jazi, M. Jadliwala, V. Namboodiri, Seer grid: privacy and utility implications of two-level load prediction in smart grids. IEEE Trans. Parallel Distrib. Syst. **28**(2), 546–557 (2017)
4. X. Chen, L. Wang, J. Ding, N. Thomas, Patient flow scheduling and capacity planning in a smart hospital environment. IEEE Access **4**, 135–148 (2016)
5. P. Xu, H. Mei, L. Ren, W. Chen, Vidx: visual diagnostics of assembly line performance in smart factories. IEEE Trans. Vis. Comput. Graph. **23**(1), 291–300 (2017)
6. M. Hassanalieragh, T. Soyata, A. Nadeau, G. Sharma, Ur-solarcap: an open source intelligent auto-wakeup solar energy harvesting system for supercapacitor-based energy buffering. IEEE Access **4**, 542–557 (2016)
7. D. Porcarelli, D. Spenza, D. Brunelli, A. Cammarano, C. Petrioli, L. Benini, Adaptive rectifier driven by power intake predictors for wind energy harvesting sensor networks. IEEE J. Emerg. Sel. Top. Power Electron. **3**(2), 471–482 (2015)
8. S. Ulukus, A. Yener, E. Erkip, O. Simeone, M. Zorzi, P. Grover, K. Huang, Energy harvesting wireless communications: a review of recent advances. IEEE J. Sel. Areas Commun. **33**(3), 360–381 (2015)

9. A.M. Z. L. M. A.S. Prabaharan, K.P. Seng, Radio frequency energy harvesting and management for wireless sensor net-works, in *Green Mobile Devices and Networks: Energy Optimization and Scavenging Techniques* (CRC Press, 2012), pp. 341–368
10. L.R. Varshney, Transporting information and energy simultaneously, in *2008 IEEE International Symposium on Information Theory* (2008), pp. 1612–1616
11. B. Koo, D. Park, Interference alignment and wireless energy transfer via antenna selection. IEEE Commun. Lett. **18**(4), 548–551 (2014)
12. L. Liu, R. Zhang, K.C. Chua, Wireless information and power transfer: a dynamic power splitting approach. IEEE Trans. Commun. **61**(9), 3990–4001 (2013)
13. Y. Dong, M.J. Hossain, J. Cheng, Joint power control and time switching for swipt systems with heterogeneous qos requirements. IEEE Commun. Lett. **20**(2), 328–331 (2016)
14. P. Grover, A. Sahai, Shannon meets tesla: wireless information and power transfer, in *2010 IEEE International Symposium on Information Theory* (2010), pp. 2363–2367
15. K. Huang, E. Larsson, Simultaneous information and power transfer for broadband wireless systems. IEEE Trans. Signal Process. **61**(23), 5972–5986 (2013)
16. X. Zhou, R. Zhang, C.K. Ho, Wireless information and power transfer in multiuser ofdm systems, in *2013 IEEE Global Communications Conference (GLOBECOM)* (2013), pp. 4092–4097
17. R. Zhang, C.K. Ho, Mimo broadcasting for simultaneous wireless information and power transfer. IEEE Trans. Wirel. Commun. **12**(5), 1989–2001 (2013)
18. H. Ju, R. Zhang, Throughput maximization in wireless powered communication networks, in *2013 IEEE Global Communications Conference (GLOBECOM)* (2013), pp. 4086–4091
19. Y.L. Che, L. Duan, R. Zhang, Spatial throughput maximization of wireless powered communication networks. IEEE J. Sel. Areas Commun. **33**(8), 1534–1548 (2015)
20. H. Ju, R. Zhang, Optimal resource allocation in full-duplex wireless-powered communication network. IEEE Trans. Commun. **62**(10), 3528–3540 (2014)
21. S. Bi, Y. Zeng, R. Zhang, Wireless powered communication networks: an overview. IEEE Wirel. Commun. **23**(2), 10–18 (2016)
22. O. Ozel, K. Shahzad, S. Ulukus, Optimal energy allocation for energy harvesting transmitters with hybrid energy storage and processing cost. IEEE Trans. Signal Process. **62**(12), 3232–3245 (2014)
23. J. Park, B. Clerckx, Joint wireless information and energy transfer with reduced feedback in mimo interference channels. IEEE J. Sel. Areas Commun. **33**(8), 1563–1577 (2015)
24. W. Sun, E.G. Strm, F. Brnnstrm, M.R. Gholami, Random broadcast based distributed consensus clock synchronization for mobile networks. IEEE Trans. Wirel. Commun. **14**(6), 3378–3389 (2015)

Chapter 4
Joint Time Allocation and User Scheduling in a Full-Duplex Aided Multi-user DEIN

Abstract In order to address the energy shortage in communication networks, radio frequency (RF) signals are exploited for transferring energy to miniature devices, which yields wireless powered communication networks (WPCNs). A full-duplex aided hybrid-base-station (H-BS) is deployed in a WPCN for simultaneously transferring energy in the downlink and receiving data in the uplink, which may significantly increase the usage efficiency of the precious radio spectrum. User equipments (UEs) may deplete all the energy received from the H-BS for powering their own uplink transmissions. Therefore, in this full-duplex WPCN, a joint time allocation and UE scheduling algorithm is proposed for the sake of maximising the sum-uplink-throughput of multiple UEs by further considering UEs' actual data uploading requirements. The numerical results demonstrate that the suboptimal solution is capable of achieving almost the same performance with its optimal counterpart, while our scheme outperforms other existing peers in terms of the sum-uplink-throughput.

Keywords Convex Optimisation · Hungarian Algorithm · Resource Allocation RF Signal Based Wireless Charging · Sum-Throughput Maximisation Time-Division-Multiple-Access (TDMA) · User Scheduling · Wireless Powered Communication Network (WPCN)

In order to address the energy shortage in communication networks, radio frequency (RF) signals are exploited for transferring energy to miniature devices, which yields wireless powered communication networks (WPCNs). A full-duplex aided hybrid-base-station (H-BS) is conceived in a WPCN for simultaneously transferring energy during downlink transmissions and receiving data during uplink transmissions. User Equipments (UEs) may deplete all the energy received from the H-BS for powering their own uplink transmissions. Therefore, in this full-duplex WPCN, a joint time allocation and UE scheduling algorithm is proposed for the sake of maximising the sum-uplink-throughput of multiple UEs by further considering UEs' actual data uploading requirements. Details will be found in this chapter and the further numerical results will also be discussed.

J. Hu and K. Yang, *Data and Energy Integrated Communication Networks*, SpringerBriefs in Computer Science, https://doi.org/10.1007/978-981-13-0116-2_4

4.1 Introduction

In order to address the energy shortage of communication devices in the 5G and IoT era, engineers are seeking renewable energy from the ambient environment for charging miniature devices in networks, which yields a growing interest in the energy harvesting technique and its employment in communication networks [1, 2]. However, harvesting renewable energy from the ambient environment has the following drawbacks: Firstly, the arrival of renewable energy heavily depends on the ambient environment, which makes the energy harvesting unreliable and uncontrollable. Secondly, harvesting renewable energy from the ambient environment requires cumbersome equipments, such as solar panels and water/wind turbines, which is not practical for supplying energy to pocket-sized communication devices.

By contrast, RF signals based wireless energy transfer has the following advantages over renewable energy harvesting: Firstly, the amplitude of RF signals can be dynamically adjusted for the sake of controlling the amount of energy arriving at energy requesters, which makes RF signals based wireless energy transfer completely controllable. Secondly, only a small-scale energy reception circuit has to be implemented in communication devices for converting the energy carried by the recieved RF signals to the useful direct-current (DC) energy, which is suitable for charging the miniature communication devices.

As a result, combining RF signals based wireless energy transfer with the conventional wireless communication technique yields wireless powered communication networks (WPCNs) [3]. Generally, the operation in WPCNs can be separated into two phases, namely the wireless energy transfer (WET) phase during the downlink transmission and the wireless information transfer (WIT) phase during the uplink transmission. In order to increase the utilisation of precious spectrum resources, full-duplex technique can be adopted by base stations (BSs) in cellular networks [4] and by relay stations in two-hop cooperative networks [5] for simultaneously realising downlink WET and uplink WIT. Resource allocation algorithms have been paid much attention for the sake of efficiently coordinating WET and WIT in WPCNs. For example, optimal power allocation scheme have been developed in [6] for balancing the performance of both the WET and WIT in relay aided two-hop communication, where the relay can consume the energy received from the source for forwarding information to the destination. Optimal relay selection scheme have been investigated in [7] for maximising the ergodic capacity of a two-hop link. Furthermore, optimal time allocation scheme have been proposed in [8, 9] for maximising the total uplink throughput for time-slotted WPCNs. Apart from the allocation of the communication resources, user scheduling is also an efficient way for improving the system performance in a multi-user scenario [10], which can be readily implemented in cellular networks since BSs can operate as central controllers. The authors of [11] have proposed a delay-aware user scheduling algorithm for the energy harvesting downlink coordinated MIMO systems.

Due to its discrete nature, user scheduling is a non-convex optimisation problem, which can not be solved in polynomial time. In order to efficiently address the user scheduling issue in delay-sensitive communication network, the complexity and the convergence rate of the user scheduling algorithms have to be taken into account. The *Hungarian* algorithm is a classic combinatorial optimisation algorithm, which is often exploited for addressing the assignment and ranking problems [12, 13]. As a result, we can rely on this algorithm for reducing the complexity of the user scheduling algorithm in WPCNs.

Unfortunately, the user scheduling issue in full-duplex WPCNs has not been well addressed. In order to fill in this gap, our novel contributions are summarised as below:

- For the purpose of maximising the sum-uplink-throughput of user equipments (UEs), we propose a joint time allocation and UE scheduling algorithm for a full-duplex WPCN by ensuring UEs' data transmission requirements.
- The UE scheduling issue in the full-duplex WPCN is modelled as a typical assignment problem, which can be addressed by exploiting the classic Hungarian algorithm.
- Furthermore, based on the UE scheduling result, both the optimal and the low-complexity suboptimal solutions are obtained for the time allocation scheme by considering the inherent relationship between the downlink WET and the uplink WIT.

The rest of this chapter is organised as follows: The preliminary knowledge of the WPCN and the full-duplex technique is introduced in Sect. 4.2. Then, the system of the full-duplex aided WPCN is described in Sect. 4.3, where both the downlink WET and the uplink WIT performance are characterised. Furthermore, both the problem formulation and the joint time allocation and UE scheduling algorithm are detailed in Sect. 4.4, while the numerical results are provided in Sect. 4.5. Finally, this chapter is concluded in Sect. 4.6.

4.2 Preliminary

4.2.1 *Wireless Powered Communication Network*

In the near-future IoT era, massive devices are connected to the network so as to make them autonomously communicate with others, think independently and make decisions intelligently. Since the IoT devices are mainly powered by the embedded batteries, the energy efficient management on the IoT devices is essential for extending the lifetime of the IoT. Nevertheless, no matter how "efficient" the energy management is, the batteries of the IoT devices have to be eventually replaced. However, the IoT devices are deployed in some untouchable places. For instance, the sensors may be implanted under the human skin in order to continuously monitor the health conditions of the patients. The sensors may also be buried in the building in order to monitor the building structure. In these circumstances, replacing the batteries for these sensors is mission impossible. We have to seek for other means for replenishing their batteries.

Thanks to the abundant RF signals in the ambient environment, the IoT devices may harvesting the RF energy for powering their communication operations. Any communication network powered by the RF energy can be regarded as wireless powered communication network. The dedicated wireless energy beacon can be deployed for powering the IoT devices. A hybrid base station can also be invoked for fulfilling both the energy and data requirement of the IoT devices. The classic protocol for the wireless powered communication network is the "harvest-then-transmit" protocol. This is a time-division protocol. In the first time slot, the IoT devices receive energy from either the dedicated wireless energy beacon or the H-BS. Then they carry out their communication tasks in the second time slot with the energy harvested during the previous time slot. The main drawback of the "harvest-then-transmit" protocol is the reduction of the throughput, since the IoT devices have to sacrifice some of the time for harvesting sufficient energy.

As a result, in this chapter, we study a full-duplex wireless powered communication network, which enables the downlink wireless energy transfer and the uplink wireless information transfer at the same time. Therefore, the time-domain resources can be reused by both the downlink wireless energy transfer and the uplink wireless information transfer. The throughput can thus be substantially increased.

4.2.2 *Full-Duplex*

In order to avoid their mutual interference, the uplink transmission and the downlink transmission are often scheduled in different time slots or in different frequency bands, which yields the time-division-duplex (TDD) and the frequency-division-duplex (FDD). However, the explosive growth of the data throughput and the massive connectivity are both predicted in the upcoming 5G and IoT era. In order to cope with these challenging requirements, we have to more efficiently exploit the limited time-frequency resources, which yields the full-duplex technique. A full-duplex aided transmitter is capable of simultaneously transmitting the information in the downlink and of receiving the information in the uplink within the same frequency band. As a result, the spectral efficiency can be significantly improved with the aid of the full-duplex technique. However, we have to suffer from the serious self-interference imposed by the downlink transmission on the uplink reception. The following pair of approaches are effective in cancelling the adverse effect of the self-interference:

- Multi-antenna aided interference suppression. The multi-antenna aided uplink receiver can be carefully designed for suppressing the self-interference imposed by the transmit antennas, while maximising the uplink throughput. This interference suppression approach acquires the knowledge of the self-interference-channel. The imperfect channel estimation may results in the residual self-interference at the receive antennas. The multi-antenna aided interference suppression is suitable for

the scenario that the users are geographically close to the full-duplex aided transmitter. As a result, the uploading signal may not be buried in the self-interference.
- Successive interference cancellation. When the user is far away from the full-duplex transmitter, their uploading signal strength is far lower than the self-interference, since the uploading signal has to traverse a long distance and it hence substantially attenuated by the long-distance signal propagation. As a result, at the receive antennas, the uploading signal may be completely buried in the self-interference. However, we may exploit the huge difference between the uploading signal and the self-interference by invoking the successive interference cancellation. Generally, the receiver may firstly decode the self-interference and remove it from the original superposition signal. Then, the uploading signal can be successfully decoded. The main advantage of the successive interference cancellation is that it does not require accurate channel state information, which reduces the receiver complexity.

In this chapter, we consider a full-duplex aided hybrid base station (H-BS), which is capable of simultaneously broadcasting energy to the user equipments (UEs) and of receiving uploaded information. However, in this work, we assume perfect self-interference cancellation. As a result, we can focus our attention on the design of the resource allocation and user scheduling algorithms.

4.3 System Model

4.3.1 Network Model

In this chapter, we consider a full-duplex WPCN, as illustrated in Fig. 4.1, which consists of a single hybrid base station (H-BS) and K user equipments (UEs). The H-BS is equipped with two independent radio fronts for the implementation of the full-duplex function. One radio front is invoked for powering the UEs by emitting RF signals during the downlink transmission, while the other radio front is invoked for receiving the information during the UEs' uplink transmissions. Note that each radio front is only equipped with a single antenna. By contrast, each UE is only equipped with a single radio front having a single antenna, which yields the half-duplex operation.

The energy reception circuits at UEs can only be activated if the power carried by the received RF signal is in the order of −20 dBm, which is much higher than the power required by information decoding. Therefore, in order to efficiently activate the energy reception circuits, UEs have to be placed within the proximity of the H-BS, which results in line-of-sight (LoS) channels between the H-BS and UEs. As a result, the uncorrelated Rician block fading is conceived for modelling the channel attenuation between the H-BS and UEs. The downlink channel power gain between the H-BS and the i-th UE is denoted as g_i, while the uplink channel power gain between this communication pair is denoted as h_i. Specifically, both the channel

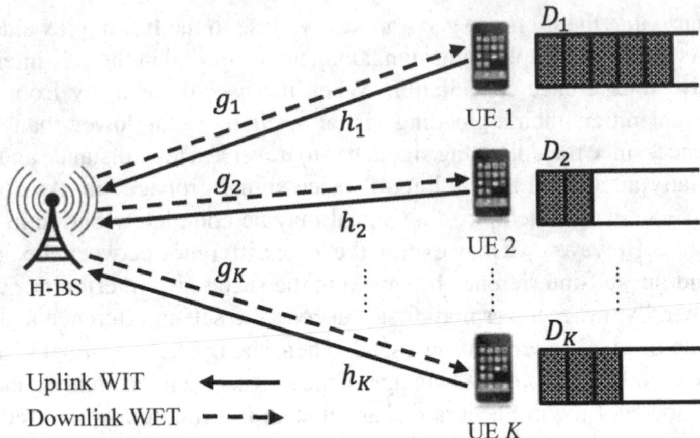

Fig. 4.1 A full-duplex aided WPCN

power gains g_i and h_i remain constant during a transmission block but vary from one transmission block to another. The perfect channel state information (CSI) can be acquired by the H-BS for the design of the optimal resource allocation and user scheduling scheme.

Considering the practical data transmission requirements of UEs, we assume that UEs have specific amount of data to be transmitted during a single transmission block. The data amount of the i-th UE willing to upload to the H-BS is denoted as D_i during a specific transmission block. Note that a UE may have different amount of data to be uploaded during different transmission blocks. As shown in the system model of Fig. 4.1, the data to be uploaded during a transmission block is stored in the buffer of UE.

4.3.2 Frame Structure

The frame structure for both the WET and WIT phases during a single transmission block in the WPCN is portrayed in Fig. 4.2. During the downlink transmission from the H-BS towards UEs, the broadcast nature of wireless channels is exploited for simultaneously transferring energy to the UEs in the WPCN. This WET phase is carried out by a single radio front of the H-BS. The downlink transmit power of the H-BS is set to be P_H. The downlink RF signal dedicated for the WET is denoted as x_H. During their uplink transmissions, UEs upload their data to the H-BS by adopting the time-division-multiple-access (TDMA) protocol for the sake of avoiding any transmission collision. The H-BS relies on another radio front for collecting data uploaded by the UEs. Since the H-BS perfectly knows the downlink RF signal x_H emitted by itself, we may reasonably assume that the self-interference imposed by the downlink WET on the uplink data reception can be effectively cancelled by some signal processing approaches.

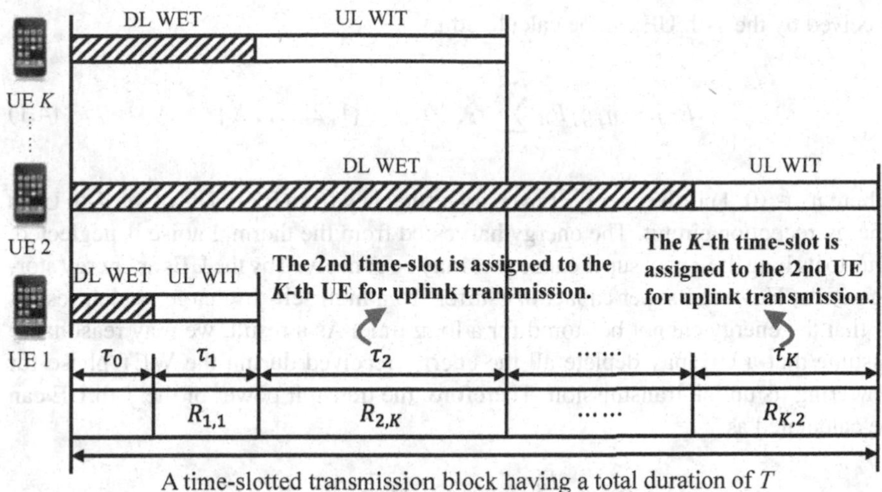

Fig. 4.2 A time-slotted transmission block for downlink WET and uplink WIT

Observe from Fig. 4.2 that the performance of a specific UE in the WPCN is affected by two key factors. The first key factor is the duration of the time slot assigned to this UE. If a longer time slot is assigned, this UE is capable of uploading more data during its uplink transmission. The second key factor is the uploading order of this UE. If the UE is scheduled in front for its uplink transmission, it may have a reduced time to harvest energy from the H-BS's downlink WET. As a result, the UE has to suffer from a throughput degradation, since its uplink transmission cannot be powered by sufficient energy. By contrast, if the UE is scheduled at back for its uplink transmission, it may have a chance to harvest more energy from the H-BS's downlink WET before its own uplink transmission. As a result, the UE may gain sufficient energy for powering its high-throughput uplink transmission.

Therefore, the prime objective of this chapter is to develop a joint time slot allocation and UE scheduling scheme for maximising the performance of the full-duplex WPCN.

4.3.3 The Downlink WET and Uplink WIT

As illustrated in Fig. 4.2, according to the basic principle of the TDMA protocol during the UEs' uplink transmissions, only a single UE is allowed to upload its data during a specific time slot. For example, considering the i-th time slot is assigned to the j-th UE during a single transmission block, the duration of this time slot is denoted as τ_i, which indicates that the j-th UE can upload its data for a duration of τ_i. Furthermore, the j-th UE is capable of receiving energy from the H-BS before it is scheduled to upload data during the i-th time slot. Hence, the total energy $E_{i,j}$

received by the j-th UE can be calculated by

$$E_{i,j} = \eta_j g_j P_H \sum_{k=0}^{i-1} \tau_k, \ \forall i, j \in \{1, 2, \ldots, K\}, \tag{4.1}$$

where $\eta_j \in (0, 1)$ is the energy conversion efficiency from RF to DC of the j-th UE's energy reception circuit. The energy harvested from the thermal noise is neglected. Due to its long life-span, super capacitors may be conceived by the UEs as energy storage units. However, super capacitors suffer from high self-discharge, which results in that the energy cannot be stored for a long time. As a result, we may reasonably assume that a UE may deplete all the energy received during the WET phase for powering its uplink transmission. Therefore, the transmit power of the j-th UE can be calculated as

$$P_{i,j} = \frac{E_{i,j}}{\tau_i}, \ \forall i, j \in \{1, 2, \ldots, K\} \tag{4.2}$$

Thanks to the TDMA protocol adopted for the medium access control, the uplink transmissions of multiple UEs may not interfere with each other, if we assume that all the UEs and the H-AP are perfectly synchronised. Then, the uplink transmission rate $R_{i,j}$ of the j-th UE during the i-th time slot, as exemplified in Fig. 4.2, can be formulated as

$$\begin{aligned} R_{i,j} &= \log\left(1 + \frac{h_j E_{i,j}}{\tau_i \sigma^2}\right) \\ &= \log\left(1 + \frac{\gamma_j \sum_{k=0}^{i-1} \tau_k}{\tau_i}\right), \forall i, j \in \{1, 2, \ldots, K\} \end{aligned} \tag{4.3}$$

where γ_j can be further expressed as $\gamma_j = \frac{h_j \eta_j g_j P_H}{\sigma^2}$ and σ^2 is the power of the Gaussian noise at the H-BS.

The UEs have various amount of data to be uploaded during a single transmission block. Since the j-th UE have D_j amount of data to upload, its actual throughput $\Gamma_{i,j}$, when it is scheduled on the i-th time slot, is expressed as

$$\begin{aligned} \Gamma_{i,j} &= \min\left(D_j, \tau_i R_{i,j}\right) \\ &= \min\left(D_j, \tau_i \log\left(1 + \frac{\gamma_j \sum_{k=0}^{i-1} \tau_k}{\tau_i}\right)\right). \end{aligned} \tag{4.4}$$

4.4 Problem Formulation

4.4.1 Sum-Throughput Maximisation

In order to schedule the uplink transmissions of K UEs in the WPCN, we divide a transmission block into K time slots, whose durations can be denoted by a vector $\boldsymbol{\tau} = [\tau_0, \tau_1, \ldots, \tau_K]^T$. Given the j-th UE's actual throughput in (4.4), the sum-uplink-throughput $\Gamma(\boldsymbol{\tau}, \mathbf{U})$ can be formulated as

$$\Gamma(\boldsymbol{\tau}, \mathbf{U}) = \sum_{i=1}^{K}\sum_{j=1}^{K} u_{i,j} \cdot \min\left(D_j, \tau_i \log\left(1 + \frac{\gamma_j \sum_{k=0}^{i-1} \tau_k}{\tau_i}\right)\right), \tag{4.5}$$

where the $u_{i,j}$ is equal to one if the j-th UE is scheduled on the i-th time slot, otherwise, zero. The K by K matrix $\mathbf{U} = \{u_{i,j}\}$ in (4.5) can be regarded as a specific UE scheduling scheme.

We aim for maximising the sum-uplink-throughput of (4.5) by finding the optimal UE scheduling scheme $\mathbf{U}^*$ and the optimal time allocation scheme $\boldsymbol{\tau}^*$, which can be formulated as

$$\text{P1: } \max_{\boldsymbol{\tau}, \mathbf{U}} \Gamma(\boldsymbol{\tau}, \mathbf{U}) \tag{4.6}$$

$$\text{s.t.: } \tau_i \geq 0, \ \forall i \in \{1, 2, \ldots, K\}, \tag{4.6a}$$

$$\sum_{i=1}^{K} \tau_i \leq 1, \tag{4.6b}$$

$$u_{ij} \in \{0, 1\}. \tag{4.6c}$$

$$\sum_{i=1}^{K} u_{ij} = 1, \ \forall j \in \{1, 2, \ldots, K\}, \tag{4.6d}$$

$$\sum_{j=1}^{K} u_{ij} = 1, \ \forall i \in \{1, 2, \ldots, K\}, \tag{4.6e}$$

4.4.2 Iterative Algorithm

For the sake of solving the sum-uplink-throughput maximisation problem P1, both the optimal and the suboptimal solutions will be found by optimising both the user scheduling scheme $\mathbf{U}$ and the time allocation scheme $\boldsymbol{\tau}$ in the following steps.

STEP 1: We initialise the time allocation scheme by equally dividing a transmission block into $(K + 1)$ time slots. Hence, the initial value of the i-th time slot's duration is expressed as

$$\tau_i^{(0)} = \frac{T}{K+1}, \forall i \in \{0, 1, 2, \ldots, K\}, \tag{4.7}$$

where T represents the total duration of a transmission block.

All the initial durations of $(K+1)$ time slots constitute the initial time slot allocation vector $\boldsymbol{\tau}^{(0)}$. We also randomly allocate a single time slot to each UE, which yields the initialisation of the UE scheduling scheme $\mathbf{U}^{(0)}$. Substituting $\boldsymbol{\tau}^{(0)}$ and $\mathbf{U}^{(0)}$ into (4.5), the initial sum-uplink-throughput $\Gamma(\boldsymbol{\tau}^{(0)}, \mathbf{U}^{(0)})$ can be calculated.

STEP 2: The order of a UE being scheduled for transmission may affect the duration of its energy harvesting operation during a transmission block. The amount of energy received by a UE may further decide its uplink transmission rate and thus its uplink throughput. Consequently, the optimisation of the time allocation scheme $\boldsymbol{\tau}$ is tangled with that of the user scheduling scheme $\mathbf{U}$. We will firstly find the optimal UE scheduling scheme in this step. Substituting $\boldsymbol{\tau}^{(0)}$ into (4.5), the resultant sum-uplink-throughput can be expressed as

$$\Gamma(\boldsymbol{\tau}^{(0)}, \mathbf{U}) = \sum_{i=1}^{K} \sum_{j=1}^{K} u_{i,j} \min\left(D_j, \frac{T}{K+1} \log\left(1 + i \cdot \gamma_j\right)\right).$$

Letting $C_{i,j} = \min\left(D_j, \frac{T}{K+1} \log\left(1 + i \cdot \gamma_j\right)\right)$, the original optimisation problem P1 can be reformulated as:

$$\text{P2}: \quad \max_{\mathbf{U}} \Gamma(\boldsymbol{\tau}^{(0)}, \mathbf{U}) = \sum_{i=1}^{K} \sum_{j=1}^{K} u_{i,j} C_{i,j} \tag{4.8}$$

$$\text{s.t.:} \quad (4.6c), (4.6d) \text{ and } (4.6e).$$

Therefore, the only variable have to be optimised in P2 is the UE scheduling scheme $\mathbf{U}$. Observe from P2 that the optimisation of the UE scheduling scheme can be regarded as a classic assignment problem, since a single time slot only allows a single UE to upload data on it and a single UE is only admitted to access a single time slot. P2 has a form of a zero-one integer programming problem. As a result, Hungarian algorithm can be exploited for finding the optimal solution $\mathbf{U}^{(1)}$. The complexity of the Hungarian algorithm aided optimisation is $\mathcal{O}(K^3)$, where the number K of UEs is also the number of edges in the maximum matching problem.

STEP 3: Substituting the solution $\mathbf{U}^{(1)}$ of the UE scheduling scheme into (4.6), P1 can be reformulated as

$$\text{P3}: \quad \max_{\boldsymbol{\tau}} \Gamma(\boldsymbol{\tau}, \mathbf{U}^{(1)}), \tag{4.9}$$

$$\text{s.t.:} \quad (4.6a) \text{ and } (4.6b).$$

Algorithm 1 Suboptimal Time Allocation Algorithm

Require: channel power gains: $\boldsymbol{g} = (g_1, \cdots, g_K)$, $\boldsymbol{h} = (h_1, \cdots, h_K)$; the energy conversion rates for UEs: $\boldsymbol{\eta} = (\eta_1, \cdots, \eta_K)$; transmission power of H-BS: P_H; the number of UEs: K; the distance between UEs and the H-BS: $\boldsymbol{d} = (d_1, \cdots, d_K)$; the amount of data to be transmitted by UEs: $\boldsymbol{D} = (D_1, \cdots, D_K)$; the initialisation of time allocation scheme $\widetilde{\boldsymbol{\tau}}^{(0)} = \boldsymbol{\tau}^{(0)}$.
Ensure: suboptimal time allocation scheme $\widetilde{\boldsymbol{\tau}}^{(1)}$;

```
Initialise Δt*_0 = 0
Initialise τ̃_0^(1) = τ̃_0^(0);
for i = 1 : K do
    if the i-UE cannot finish its transmission during τ_i + Δt(i − 1) then
        Δt(i) = 0;
    else
        Δt*_i = arg max Γ^ach_{i+1,i+1}, where Γ^ach_{i+1,i+1} is the achievable throughput of the (i + 1)-th UE, as formulated in (4.10);
    end if
    τ̃_i^(1) = Δt*_{i−1} + τ̃_i^(0) − Δt*_i
end for
return τ̃^(1) = {τ̃_0^(1), τ̃_1^(1), ···, τ̃_k^(1)};
```

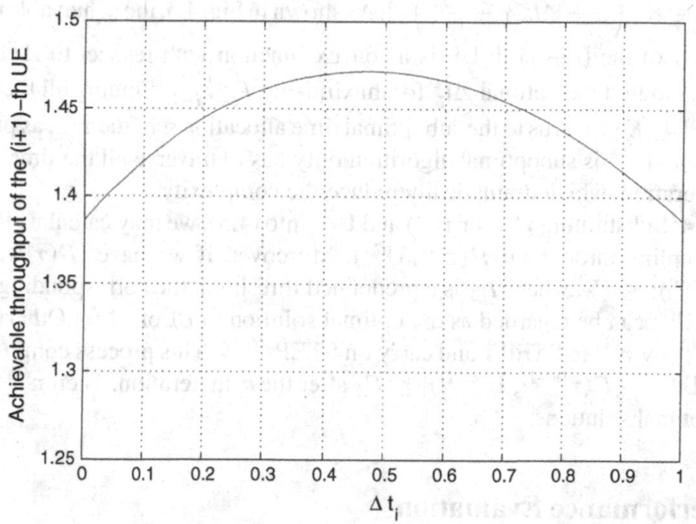

Fig. 4.3 The achievable throughput $\Gamma^{ach}_{i+1,i+1}$ of the $(i+1)$-th UE with respect to $\Delta t(i)$

According to its formulation (4.5), the sum-uplink-throughput is convex with respect to $\boldsymbol{\tau}$, when the UE scheduling scheme $\mathbf{U}$ is fixed. Furthermore, since (4.6a) and (4.6b) are both affine, P3 is a convex optimisation problem. Hence, the optimal solution $\boldsymbol{\tau}^{(1)}$ can be obtained by exploiting convex optimisation tools.

In order to further reduce the complexity of the P3's optimisation process, a suboptimal solution can be obtained by carrying out Algorithm 1. Since some UEs may have a small amount of data to be uploaded, they may finish their uplink transmissions before the end of their assigned time slots, which results in the time slots under full exploitation. The basic idea of Algorithm 1 is to re-allocate the redundant time to the other UEs, who fail to finish their uplink transmission during their assigned time slots.

As depicted in Algorithm 1, the redundant time Δt_i during which the i-th UE does not actually transmit any data is reallocated to the $(i+1)$-th UE. Then, the duration of the $(i+1)$-th UE's time slot can be recalculated as $(\tau_{i+1} + \Delta t_i)$. If the $(i+1)$-th UE can finish its transmission before the end of its time slots having the recalculated duration, the redundant time may be delivered to the next UE. If the $(i+1)$-th UE cannot finish its transmission during this renewed time slot, the duration of the next UE's time slot do not have to be renewed. Without considering its data requirement D_{i+1}, the achievable throughput of the $(i+1)$-th UE[1] can be formulated as

$$\Gamma^{ach}_{i+1,i+1} = (\Delta t_i + \tau_{i+1}) \cdot \log\left(1 + \frac{\gamma_j \sum_{k=0}^{i-1} \tau_k}{\Delta t_i + \tau_i}\right), \tag{4.10}$$

where $\Delta t_i \in [0, \left(\tau_i + \Delta t_{i-1} - \frac{D_i}{\Gamma^{ach}_{i,i}}\right)^+]$. As shown in Fig. 4.3, the achievable throughput $\Gamma^{ach}_{i+1,i+1}$ of the $(i+1)$-th UE is a convex function with respect to Δt_i. Hence, we are able to find the optimal Δt_i^* for maximising $\Gamma^{ach}_{i+1,i+1}$. Finding all the optimal $\{\Delta t_i^* | 0 \le i \ge K\}$ leads us to the suboptimal time allocation scheme $\widetilde{\boldsymbol{\tau}}^{(1)}$, as presented in Algorithm 1. This suboptimal algorithm only has to traverse all the time slots for a single iteration, which dramatically reduce the complexity.

STEP 4: Substituting $\boldsymbol{\tau}^{(1)}$ (or $\widetilde{\boldsymbol{\tau}}^{(1)}$) and $\mathbf{U}^{(1)}$ into (4.5), we may calculate the resultant sum-uplink-throughput $\Gamma(\boldsymbol{\tau}^{(1)}, \mathbf{U}^{(1)})$. Moreover, if we have $|\Gamma(\boldsymbol{\tau}^{(1)}, \mathbf{U}^{(1)}) - \Gamma(\boldsymbol{\tau}^{(0)}, \mathbf{U}^{(0)})| \le \Gamma_\Delta$, where Γ_Δ is a predefined threshold, the corresponding $\boldsymbol{\tau}^{(1)}$ (or $\widetilde{\boldsymbol{\tau}}^{(1)}$) and $\mathbf{U}^{(1)}$ can be regarded as the optimal solution to P1 of (4.6). Otherwise, we replace $\boldsymbol{\tau}^{(0)}$ by $\boldsymbol{\tau}^{(1)}$ in STEP 2 and carry on STEPs 3–4. This process continues until $|\Gamma(\boldsymbol{\tau}^{(n)}, \mathbf{U}^{(n)}) - \Gamma(\boldsymbol{\tau}^{(n-1)}, \mathbf{U}^{(n-1)})| \le \Gamma_\Delta$ after the n-th iteration. Then $\boldsymbol{\tau}^{(n)}$ and $\mathbf{U}^{(n)}$ are the optimal solutions.

4.5 Performance Evaluation

In this section, numerical results are provided for demonstrating the advantages of our proposed joint time allocation and user scheduling algorithm over other existing counterparts in terms of the sum-uplink-throughput in WPCNs.

For simplicity, the Gaussian noise power σ^2 is assumed to be a unity. The number of UEs in the WPCN is $K = 10$. The RF-DC energy conversion efficiencies for all the K UEs can be denoted by a set $\boldsymbol{\eta} = \{\eta_k \in (0, 1] | k = 1, 2, \ldots, K\}$. The distances from the H-BS to UEs are uniformly chosen from 1 to 10 m. As a result, LoS paths from the H-BS to UEs exists. Hence, both the downlink and uplink propagations

[1] After the UE scheduling, the index of the UE, which is scheduled on the $(i+1)$-th time slot, can also be denoted as $(i+1)$.

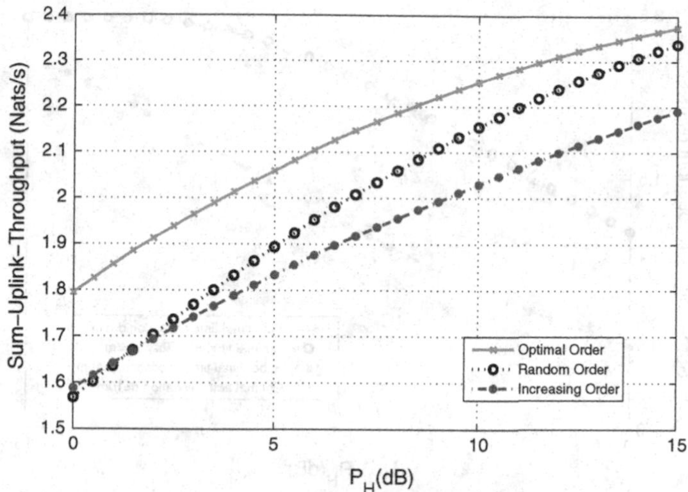

Fig. 4.4 Sum-uplink-throughput of different UE scheduling schemes. All the UE scheduling schemes are associated with the optimal time allocation as proposed in Sect. 4.3

are attenuated by the Rician fading. Furthermore, the channel power gains remain constant during a single transmission block but vary from one block to another.

We plot the sum-uplink-throughput of three different UE scheduling schemes in Fig. 4.4. According to [8], we can achieve a higher sum-uplink-throughput when the UEs are ranked for uploading their data in an increasing order of their signal-to-noise-ratios (SNRs) than when the UEs are ranked in an decreasing order of their SNRs. However, this result is not convincing since the sizes of the UEs' uploaded

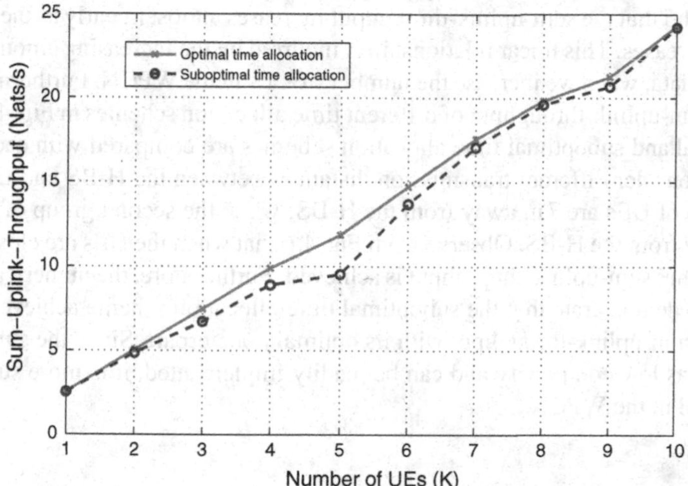

Fig. 4.5 Sum-uplink-throughput versus the number of UEs

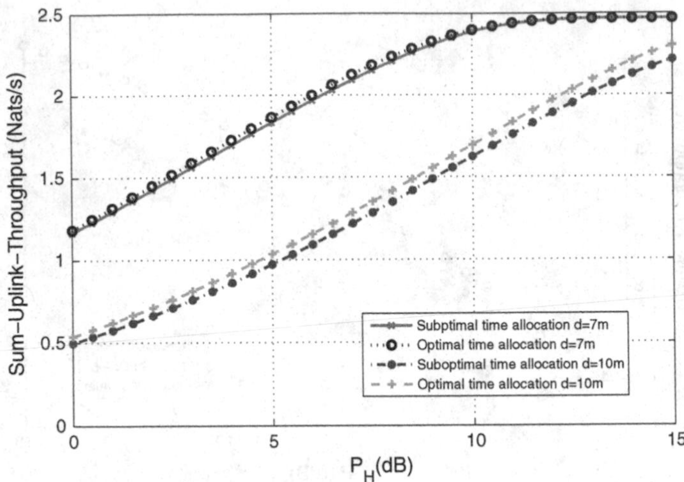

Fig. 4.6 Sum-uplink-throughput of different time allocation schemes

data are not considered. For instance, a UE having a good channel condition may not have much data to be uploaded. As a result, scheduling the UEs corresponding to their increasing order of SNRs may waste some precious time resources. We may observe from Fig. 4.4 that our optimal UE scheduling scheme outperforms its counterparts. Furthermore, as shown in Fig. 4.4, the sum-uplink-throughput grows quickly at first and then converges to a constant as we increases the transmit power of the H-BS. This is because the finite data uploaded by the UEs limits the growth of the sum-uplink-throughput.

We plot the sum-uplink-throughput versus the number of UEs in Fig. 4.5. Observe from Fig. 4.5 that the sum-uplink-throughput increases almost linearly as the number of UEs increases. This linear relationship is incurred by the increasing amount of the uploaded data, when we increase the number of UEs in the WPCN. Furthermore, we plot the sum-uplink-throughput of different time allocation schemes in Fig. 4.6. Both the optimal and suboptimal time allocation schemes are compared with each other. We also consider different transmission distances between the H-BS and UEs. The first group of UEs are 7 m away from the H-BS, while the second group of UEs are 10 m away from the H-BS. Observe from Fig. 4.6 that when the UEs are closer to the H-BS, higher sum-uplink-throughput is achieved. Furthermore, the numerical results of Fig. 4.6 demonstrate that the suboptimal time allocation scheme achieves almost the same sum-uplink-throughput with its optimal counterpart. Since the suboptimal solution has low complexity and can be readily implemented, it is more suitable to be adopted in the WPCN.

4.6 Summary

In this chapter, we study a joint time allocation and UE scheduling algorithm for maximising the sum-uplink-throughput of the UEs in a full-duplex aided WPCN, where the UEs receive the energy from the H-BS's downlink transmission and they deplete all the received energy for supporting their uplink transmissions. By considering various data uploading requirements of the UEs, the sum-uplink-throughput maximisation problem can be addressed by iteratively carrying out Hungarian algorithm aided user scheduling and by optimising the time allocation among the UEs. The numerical results demonstrate the advantage of our proposed algorithm over other existing counterparts, which may shed light on the medium access control design of the WPCN.

References

1. H. Ju, R. Zhang, Throughput maximization in wireless powered communication networks, in *2013 IEEE Global Communications Conference (GLOBECOM)* (2013), pp. 4086–4091
2. R. Xie, F.R. Yu, H. Ji, Y. Li, Energy-efficient resource allocation for heterogeneous cognitive radio networks with femtocells. IEEE Trans. Wirel. Commun. **11**(11), 3910–3920 (2012)
3. K. Yang, Q. Yu, S. Leng, B. Fan, F. Wu, Data and energy integrated communication networks for wireless big data. IEEE Access **4**, 713–723 (2016)
4. H. Ju, E. Oh, D. Hong, Improving efficiency of resource usage in two-hop full duplex relay systems based on resource sharing and interference cancellation. IEEE Trans. Wirel. Commun. **8**(8), 3933–3938 (2009)
5. D. Nguyen, L.N. Tran, P. Pirinen, M. Latva-aho, Precoding for full duplex multiuser MIMO systems: spectral and energy efficiency maximization. IEEE Trans. Signal Process. **61**(16), 4038–4050 (2013)
6. Z. Zheng, L. Song, D. Niyato, Z. Han, Resource allocation in wireless powered relay networks: a bargaining game approach. IEEE Trans. Veh. Technol. **PP**(99), 1 (2016)
7. D.S. Michalopoulos, H.A. Suraweera, R. Schober, Relay selection for simultaneous information transmission and wireless energy transfer: a tradeoff perspective. IEEE J. Sel. Areas Commun. **33**(8), 1578–1594 (2015)
8. X. Kang, C.K. Ho, S. Sun, Full-duplex wireless-powered communication network with energy causality. IEEE Trans. Wirel. Commun. **14**(10), 5539–5551 (2015)
9. H. Ju, R. Zhang, Optimal resource allocation in full-duplex wireless-powered communication network. IEEE Trans. Commun. **62**(10), 3528–3540 (2014)
10. X. Xia, G. Wu, S. Fang, S. Li, SINR or SLNR: in successive user scheduling in MU-MIMO broadcast channel with finite rate feedback, in *2010 International Conference on Communications and Mobile Computing (CMC)*, vol. 2 (2010), pp. 383–387
11. Y. Cui, V.K.N. Lau, Y. Wu, Delay-aware BS discontinuous transmission control and user scheduling for energy harvesting downlink coordinated mimo systems. IEEE Trans. Signal Process. **60**(7), 3786–3795 (2012)
12. H.W. Kuhn, The Hungarian method for the assignment problem. Naval Res. Logist. (NRL) **52**(1), 7–21 (2005). https://doi.org/10.1002/nav.20053
13. Harold W. Kuhn, Statement for naval research logistics. Naval Res. Logist. (NRL) **52**(1), 6 (2005). https://doi.org/10.1002/nav.20057

Chapter 5
Conclusions and Open Challenges

Abstract We conclude this book by summarising the main contributions and results of Chaps. 1–4 concerning the information theoretical essence, transceiver architecture, resource allocation of the integrated data and energy communication network (DEIN). However, we are still facing the following critical challenges in the DEIN. The hardware of the transceiver should be further improved in order to increase the wireless energy transfer (WET) efficiency. The heterogeneous nature of diverse WET techniques should be exploited for realising the future Internet of Energy (IoE). The performance limit of the WET should be explored from the information theoretical perspective. The interference imposed by the WET signal on the WIT signal should be properly cancelled in the multi-user DEIN. The DEIN aided mobile cloud computing is another promising application, which is worth intensive efforts.

Keywords Heterogeneous Internet of Energy · Interference Cancellation · Mobile Cloud Computing · Signal Decoupling · Social-Aware Design · Wireless Energy Transfer Efficiency · Wireless Energy Transfer Capacity

In this chapter, the main conclusions of our brief are summarised and the open research challenges are reviewed.

5.1 Conclusions

In order to address the energy shortage of electronic devices in the modern wireless communication network, the RF signals are relied upon for delivering the energy to the far field. However, coordinating the wireless energy transfer (WET) and the wireless information transfer (WIT) in both the RF spectral band imposes great challenges on the system design, which yields the emerging research trend of the data and energy integrated communication network (DEIN). This brief aims for providing the landscape picture of the DEIN, while presenting the latest research in both the physical layer and the medium access control layer. Specifically, the main conclusions are drawn as follows:

J. Hu and K. Yang, *Data and Energy Integrated Communication Networks*, SpringerBriefs in Computer Science, https://doi.org/10.1007/978-981-13-0116-2_5

- In Chap. 1, the overview of data and energy integrated communication networks (DEINs) is provided. We look into the energy supply issue of the electronic devices, compare the popular wireless charging techniques with one another and highlight the RF signal based WET and its distinctive features against the conventional wireless communication in the same spectral bands. At last, the ubiquitous architecture of DEINs is portrayed by synthesising a diverse range of WET and WIT scenarios into it.
- In Chap. 2, we firstly overview the information theoretical fundamental of the integrated data and energy transfer from the perspective of the discrete-input-discrete-output memoryless channel and the continuous-input-continuous-output channel. Key enabling modules of the generic transceiver architecture for the integrated WET and WIT are introduced. Furthermore, we focus our attention on the signal splitter based receiver architecture equipped with multiple antennas, which includes the spatial splitting based receiver, power splitting based receiver and the time switching based receiver. Finally, the rate-energy region of the signal splitter based receivers is evaluated. The numerical result tell us that the power splitting based receiver outperforms its time switching based counterpart, when the linear RF-DC converter is conceived.
- In Chap. 3, we have studied a novel DEIN model, where the hybrid base station (H-BS) simultaneously transmits the data and energy during the downlink transmissions and the UEs harvest the energy from the downlink signals for powering their own uplink transmissions. In order to reduce potential collision and interference, a time-division-multiple-access (TDMA) protocol is adopted in the medium-access-control layer for both the downlink and uplink transmissions. At a UE's end, the received RF signal is split in the power domain. One portion of the signal is for information decoding, while the other is for energy harvesting. Relying on the classic convex optimization theory, both of the sum-throughput and the fair-throughput are maximised by optimizing both of the time slots allocation and the power spliting factors. Iterative algorithms are proposed for numerically solving the throughput maximization problems. Furthermore, our numerical results demonstrate the advantage of our DEIN over the WPCN and the WIT systems.
- In Chap. 4, we study a joint time allocation and UE scheduling algorithm for maximising the sum-uplink-throughput of the UEs in a full-duplex aided WPCN, where the UEs receive the energy from the H-BS's downlink transmission and they deplete all the received energy for supporting their uplink transmissions. By considering various data uploading requirements of the UEs, the sum-uplink-throughput maximisation problem can be addressed by iteratively carrying out Hungarian algorithm aided user scheduling and by optimising the time allocation among the UEs. The numerical results demonstrate the advantage of our proposed algorithm over other existing counterparts, which may shed light on the medium access control design of the WPCN.

5.2 Open Challenges

There are numerous open challenges in the research of DEINs, which are briefly summarised in this section.

5.2.1 *Efficiency Enhancement of WET*

Although RF signals are capable of carrying remarkable energy to far-field locations, the WET efficiency remains low due to the serious channel attenuation incurred by path-loss, shadowing and fast fading. As a result, a jointly optimised design has to be carried out by considering the entire WET chain, including the DC-RF energy converter at the transmitter, transmit/receive antenna design, the RF signal propagation over wireless channels and the RF-DC energy converter at the receiver. For instance, solid-state distributed RF amplifiers have to be invoked for the sake of increasing the DC-RF energy conversion efficiency at the transmitters. Beamforming relying on holographic antennas should also be implemented in order to counteract the serious channel attenuation. Furthermore, non-diffraction beams are capable of reducing the energy loss during the RF signals' propagation. Therefore, exploiting non-diffraction beams may be capable of remarkably increasing both the WET and the WIT performance. Additionally, the powerful meta-surface based energy harvesting technique should be implemented at the receivers in order to efficiently glean energy from the received RF signals, while advanced RF-DC circuits and battery charging circuits also have to be carefully designed.

Furthermore, the mathematical model of evaluating the efficiency of the entire WET chain remains an open problem in the literature, which requires joint efforts from both the electronic and communication engineers. This model may also be relied upon for jointly optimising all the components in a WET chain.

5.2.2 *Efficient Energy Storage*

In DEINs, the heterogeneous UEs may have different energy requirements, similar to their WIT counterparts. For example, some UEs require fast charging, which can be satisfied by establishing narrow energy beams gleaned from the energy beacons. Therefore, extremely high energy bursts may be delivered to the UEs in a very short instant. In order to accommodate these energy peaks, we have to focus our attention on the design of energy storage and management schemes from the perspective of battery energy density. Furthermore, more efforts should also be invested in charging/discharging management, energy balancing management and thermal management.

In DEINs, some UEs only require micro-level energy for supporting their low-power operations, which can be satisfied by establishing wide energy beams spanning from the energy beacons to the multiple energy requesters. Since the energy is distributed across a wide area, only a small fraction of the energy can be delivered to a single energy requester. Sometimes, the energy carried by the RF signals is too low to efficiently charge batteries. As a result, advanced voltage multiplier techniques have to be explored in order to efficiently collect low-levels of energy in the batteries. Additionally, the management system for the collection and storage of low-level energy all require further attention, as an important part of energy charging techniques.

5.2.3 Heterogeneous Internet of Energy

Integrating all sorts of WET techniques together, including both near-field and far-field ones, may indeed satisfy the energy requirements of UEs. However, as stated in Sect. 1.2 of Chap. 1, different WET techniques exhibit different charging ranges and charging efficiencies. As a result, coordinating these distinctive WET techniques and exploiting their cooperation is essential for further increasing the attainable WET efficiency. By considering the specific energy levels of the sources and the energy requirements of the requesters, both the energy beacons centralised and the UEs centralised networking and association techniques can be investigated by borrowing the classic geometric stochastic theory from the research of established wireless communication.

Furthermore, upon considering the distinctive features of WET, we believe that a maximum of two-hop WET can be realised for coordinating the energy flows in the heterogeneous Internet of Energy. For example, laser based WET is capable of realising high-efficiency point-to-point energy transmission. An energy relay station is capable of receiving a substantial amount of energy carried by laser. This energy may be further multicast to low-power devices in a wide RF beam. As a result, the energy relay stations have to be equipped with heterogeneous energy transceivers. Furthermore, mobile energy stations can also be invoked for delivering energy from the source to the destinations. Fully exploiting the features of different WET techniques is capable of significantly extending the coverage of the Internet of Energy.

Moreover, efficient communication networking is also required for pilot exchange between the energy beacons and the energy requesters in order to promptly establish energy links for WET. As a result, the implementation of heterogeneous Internet of Energy has to rely on the integration with WIT, which augments the significance of DEIN again.

5.2.4 *Information Theoretic WET Capacity*

As stated in Sect. 2.1 of Chap. 2, the rate-energy function is obtained by considering the energy required by the receivers as constraints in order to maximise the mutual information between the transmitters and receivers. This methodology inherently regards WET as a by-product of conventional WIT in DEINs. However, both WET and WIT in DEINs should be fairly considered. As a result, designers of DEINs are also interested in the performance limits of WET in DEINs. Therefore, by following a similar methodology to that of Sect. 2.1 of Chap. 2, we can change the objective function to the maximisation of the energy arriving at the receivers, which is subject to the mutual information requirement. Furthermore, the methodology can be extended for studying the performance limits of the integrated data and energy transfer in broadcast channels, in MIMO channels and in interference channels, just to name a few.

5.2.5 *Interference Cancellation and Signal Decoupling*

As stated in Sect. 1.4 of Chap. 1, WET and WIT requires different levels of input power in order to activate the energy reception circuits and the information reception circuits. The substantial gap between the input power requirements imposes challenges on the integrated data and energy transfer. DEIN stations are equipped with both functions of WET and WIT. DEIN UEs may receive both energy and data from same sources. In order to counteract the adverse effects of path-loss and hence to increase amount of energy received, in one hand, DEIN stations have to increase their transmit power, or the DEIN UEs have to be closer to the DEIN stations. If DEIN UEs are distant from the DEIN stations, the RF signals arrive at a low power. These low-power RF signals can only be exploited for data reception, but they are not capable of activating the energy reception circuits. In a nutshell, DEIN UEs are only capable of receiving energy in the near-field of DEIN stations. By contrast, when DEIN UEs are in the far-field of DEIN stations, they may only receive data. This is the near-far effect in DEINs.

A potential solution for overcoming the near-far effect is to deploy a separated dedicated WET and WIT infrastructure, respectively. Specifically, WET stations should have a high transmit power and they should be densely deployed within the proximity of DEIN UEs in order to reduce the energy loss incurred by path-loss. By contrast, WIT stations could be deployed in the far-field. Due to the channel attenuation, the received power of the RF signals carrying data is much lower than that carrying energy. The RF signals dedicated to WET may impose destructive interference on those dedicated to WIT. As a result, further efforts should be invested in improving interference cancellation and signal decoupling.

5.2.6 Socially Aware Placement of DEIN Stations

In practical scenarios, DEIN UEs are unevenly distributed. They have their own favourite places to visit and some places of interest may attract many DEIN UEs. As a result, when we deploy DEIN in a specific area, we have to firstly understand the UEs' geographic preferences. We may model the UEs' movement according to their geographic preferences and study the steady-state probability of UEs being at a specific position. Different deployment schemes of DEIN stations may be designed for serving different purposes. For example, deployment schemes should be designed for maximising the successful data delivery rate, or for maximising the energy reception rate. Since both data and energy are transferred by RF signals, we should carefully design our schemes for balancing both the WET and WIT.

5.2.7 DEIN Aided Mobile Cloud Computing

Mobile cloud computing can also be integrated with our DEINs in order to support effective computation in low-power and low-complexity IoT devices. Specifically, a DEIN station may transfer energy to IoT devices for the sake of supporting their local computation or for supporting the computation to be uploaded to a node having better computing resources. Moreover, a DEIN station is also capable of offloading computing tasks from IoT devices to the mobile cloud. A set of policies has to be optimised for energy-efficient computing. First, the computing policy has to be optimised concerning whether the IoT devices should upload their computing tasks to the cloud or they should locally process them. Secondly, the IoT devices should optimally choose the time sharing between the WET phase and the computing/uploading phase so that the IoT devices may receive sufficient energy for supporting their subsequent operating phase. Furthermore, both the CPU-cycle statistics and the channel state information should be taken into account for maximising the computing related performance.

Index

J. Hu and K. Yang, *Data and Energy Integrated Communication Networks*, SpringerBriefs in Computer Science, https://doi.org/10.1007/978-981-13-0116-2

Printed in the United States
By Bookmasters